PEOPLEWORK

PeopleWork

THE HUMAN TOUCH *in* WORKPLACE SAFETY

It's time to stop playing the game of enforcement and
move safety from compliance to community.

KEVIN BURNS

PEOPLEWORK

The Human Touch in Workplace Safety

ISBN   978-1-61961-523-6 *Paperback*

978-1-61961-524-3 *Ebook*

LIONCREST
PUBLISHING

*To Trish, whose insight
inspired this book.*

CONTENTS

INTRODUCTION: GETTING PERSONAL 11

1. SET ASIDE THE RULE BOOK....................................23

2. THE CASE OF THE MISSING P.............................. 49

3. MANAGEMENT: RELATIONSHIPS MATTER.........................59

4. MANAGEMENT: PEOPLE VIEW................................ 81

5. MEETINGS: SPINNING THE REQUIREMENTS........................95

6. MEETINGS: CREATING EMPLOYEE BUY-IN...........................111

7. MARKETING: SELLING SAFETY 129

8. MARKETING: HOW TO BUILD A SUCCESSFUL SAFETY MARKETING CAMPAIGN 143

9. MOTIVATION: SUPERVISORS AND SAFETY PEOPLE...........161

10. MOTIVATION: HOW TO INFLUENCE YOUR EMPLOYEES....171

11. LAUNCH: BECOME A SAFETY LEADER.................................189

ABOUT THE AUTHOR 211

Our prime purpose in life is to help others. And if you can't help them, at least don't hurt them.

DALAI LAMA

INTRODUCTION: GETTING PERSONAL

Safety is personal. Anyone who works with heavy machinery, or at the bottom of a mine shaft, or in the oilfields already knows it in their bones. Nothing is more personal than losing your livelihood to injury, and no relationship, except for family, is tighter than among working crews. When you're on a construction scaffold or suspended over an open pit, every move everyone makes impacts you personally.

If you can't trust the guy next to you on a work site, one of you shouldn't be there.

I was already earning my living as a safety management consultant before I began to take my own job personally. It took a woman I hadn't seen in thirty years to open my eyes.

Trish had been the smartest and prettiest girl in my fourth grade class, but our time together was short-lived. At the end of the school year, she skipped a grade, which left me languishing a grade behind the girl of my dreams. By fifteen, she was prettier than ever and way out of my league, but I got up the nerve to ask her to the high school dance. When she said yes, I was shocked. I figured she was just being nice. We had fun together, but I didn't have the guts to ask her for another date.

After high school, Trish went to college, and I went into the far reaches of Northern Ontario to pursue work. Thirty years went by. By then, I was living in Calgary about two thousand miles away from my hometown when I heard about a high school reunion. I went back not knowing what to expect, but when I walked into my old high school, there she was, that pretty girl. My mad high school crush reawakened, and eight months later, we were living together in Calgary. So I'm still dating way over my head.

Not long after we got together, Trish told me, "It's taken thirty years for us to find each other. Since we're getting such a late start, I really want you to be around for a long, long time." Wow. It was probably the most meaningful thing I'd ever heard. I wish that everyone could know what it feels like to hear those words.

What she was driving at was my cigarette smoking. She was concerned for my health, and I knew I had to do something. I couldn't disappoint the girl who owned my heart.

Trish's words and the feeling behind them changed my whole way of seeing things. My entire viewpoint on life and work began to take on new meaning. Real caring and concern for another's well-being became my basis for a new way of thinking about workplace safety. Suddenly, everything made sense to me, and I began to develop a method for personalizing safety in the workplace.

THE SAFETY PROBLEM

At no time in history have there been better processes and procedures in workplace safety, and at no time in history have there been more certified safety professionals. At the same time, the number of workplace incidents keeps rising across the board. Forty-five years of workplace safety laws have somehow failed to make progress in stopping people from getting hurt at work.

When I looked at these facts, what I finally came to realize was that workplace safety hadn't yet gotten personal. It has always been promoted as a set of rules, procedures, and processes to comply with safety laws. Yet rules and regulations only answer the what, how, when, and where of safety. There's no answer for why, except, perhaps, "So that no one gets hurt."

However, people are still getting hurt. In fact, they're getting hurt in larger numbers, so safety is clearly not working as well as it could. More processes and procedures heaped on people who aren't listening, who aren't engaged and fully present, just isn't working.

What I came to understand is that we have to stop treating safety as a bunch of rules to enforce. It's time to shift the focus. Rules don't really motivate people. We need to rethink workplace safety as a personal value. Safety's got to be able to reach people at a level that's equal to the loving whispers of someone who matters—someone who says, "I want you around for a long, long time."

In my work as a safety communications consultant, I've seen that when frontline supervisors buy into safety as a personal value, they better understand their role in keeping the workplace safe. In fact, if crews themselves can become safety leaders, the need for safety inspectors would probably disappear altogether.

SAFETY IS ALL ABOUT RELATIONSHIPS

This book introduces the next level in safety. Workplace

safety lies in the relationship between the frontline employee, the employee's immediate supervisor, and the bond among the entire crew. It's where the rubber meets the road. No senior-level initiatives, safety department compliance measures, or culture improvement ideas can have positive results if the frontline supervisor hasn't established real working relationships. The supervisor needs to be a tipping point between safety compliance and safety success.

Supervisors are uniquely positioned to bring workplace safety past compliance and across the threshold to where safety is personal. When trust and respect are embraced by frontline supervisors, their ability to personally influence frontline employees is deeply improved.

All the certified safety officers, vice presidents, and safety managers in the world can't have the impact on safety that a single supervisor and a solid crew can. A rules-based approach to management doesn't have the reach of smart coaching and mentoring for ensuring safety. Quoting the rule book, finding fault, and barking orders aren't leadership. It isn't even good management.

The more I thought about workplace safety, what I began to see clearly was that the relationship between a frontline supervisor and a frontline employee is critical to the health and safety of an organization. It's where the culture of an organization is made and reinforced. Organizations thrive at the level of teamwork, camaraderie, solid work ethic, and values. A good supervisor will keep a team together, while a poor supervisor will turn over staff. We all know people who've left jobs because of a lousy boss, even at good companies.

There's no way to bleed money faster than to have a stop-work order because of an injury. The best production is safe production. That's where the highest profit is. Rushing doesn't get the job done; it only forces crews to make mistakes.

The fact is that most frontline supervisors ascend to their positions by virtue of being the most senior guy on the job. They don't have any particular management or supervisory skills, yet they're the ones in charge on-site. They're the ones who are supposed to keep the team together, keep them motivated and focused, make the right decisions, keep their crews safe.

To get safety right, they need to be armed with more than just a rule book of procedures. No one wants a safety cop looking over their shoulder while they work. Supervisors need the personal skills to become centers of influence. They need to be thinking, "What can I bring to the job that helps the new guy from getting hurt or getting others hurt? What can I do to help him or her improve?"

A title of manager or supervisor doesn't make someone a safety leader. Safety officers with college degrees to enforce Occupational Safety and Health regulations aren't necessarily safety leaders. They may be on a work site with a clipboard observing all the details, but they haven't earned the distinction of safety leader—not yet.

A safety leader takes safety beyond compliance to a shared goal that recognizes the importance of each individual on the job. As you'll see in this book, past all the processes and procedures are people. Ultimately, safety resides in the value we place on individuals and the interaction of teams. It's about how we connect and interact on the work site, how we communicate, and the value we place on each other—which is why I call this book, *PeopleWork*—because safety starts with people.

LOOKING FOR UNSAFETY

Most supervisors and safety people start in the wrong place. They go out into the field, but they don't look for safety. They look for unsafety—anything that's unsafe—whether it's people, things, or situations. In other words, the whole safety industry has been built on the wrong premise. There's been more focus on what not to do on a work site than on what to do.

Workplace safety has become a long, slow process of one-by-one elimination of things that are unsafe, which is unfortunate, because that long list of what not to do often has no direct correlation to the right actions to take.

It's like going to the grocery store with a list of items not to buy. You have to cross-reference each item against your list. What should take a half hour can take all day. It's not practical, and, worse, it's self-defeating. Everyone knows it's better to have a list of what to buy.

For example, when companies hire, they look for employees with the most positives in their favor. They don't look for people who have the fewest negatives. They choose staff, supervisors, managers, and senior executives based on all the things that are right about them. A company wants to build on its strengths, knowing full well that by building on those strengths, it will eventually eliminate its weaknesses, or at least compensate for them.

This same focus should carry over to the workplace itself. Workplaces are where decisions are made. New ideas and technology constantly disrupt old plans and strategies. Companies, therefore, want people who can think on their feet. They may not know it, at least not yet, but when it comes to safety, senior management wants people who can reach the next level in making workplaces safe.

Companies need frontline supervisors with the skills and traits to better influence frontline teams. They need supervisors who can be welcomed as influencers, people who can change the safety discussion from rule based to performance based.

Unfortunately, most supervisors aren't looking for opportunities to mentor and influence others on safety. They work in industries where safety is a long drawn-out process of finding unsafety and fixing it. However, fast-paced, changing workplaces make this method impractical.

Instead, it's time to focus on what to do.

Human resources experts talk about improving employee engagement to boost morale and productivity. Engagement results when staff feel connected to their work and can find ways to excel. Likewise, workplace safety is most effective when it builds upon communication and the human element. Supervisors who coach and mentor for better performance are more likely to get it. Coaching takes dedication. To criticize, all you need is a rule book.

THE NEW SAFETY MODEL

It's time to make safety personal. In *PeopleWork*, I present a working system for how supervisors, crews, managers, and safety people can change the results in workplace safety by first changing their own perceptions and their willingness to get personal.

Workplace safety is closely aligned with how we connect, communicate, and interact with each other on the job. It's about the value we place on each other as workmates and individuals.

When you care about the person working next to you, every action you take will be influenced by safety. When your actions on the job are designed to get the job done safely and to protect yourself and those around you, at that point you've become a safety leader.

In the work world, it's the crew supervisor who sets the standard. When the supervisor embraces a safety dynamic built around shared responsibility and real caring, the entire team performance changes. I've seen that when supervisors take safety personally and put the new safety model into play, frontline workers get the message.

Most young people in the trades today were raised in a culture where relationships matter. They're the Oprah generation, and so the human element in work comes naturally to them. They are ready to respond to a supervisor who values teamwork, camaraderie, and working relationships that promote the new culture of safety.

For frontline supervisors, *PeopleWork* lays out a new safety model. It changes the discussion from rule-based enforcement to performance-based culture focused on mentoring, coaching, and inspiring teams.

In these pages, you will find out how to positively influence safety outcomes by influencing your crews. You'll learn practical methods for building a people-based approach to safety. You'll see how support systems build trust and motivation. You'll learn communication strategies for motivating your crews.

Most importantly, you'll learn how to take workplace safety to a level where it actually works. Along the way, you'll become a safety leader, and if you're really good, so will your crew.

CHAPTER

SET ASIDE THE RULE BOOK

Every job serves someone else. It's a point that I make at international conferences of safety professionals and at safety meetings of frontline employees when I'm invited to speak. Whether you're a general laborer and it's your first day on the job or you're a senior management executive with more letters behind your name than in your name, your job is ultimately service—to customers, to the company, and, most importantly, to each other.

It's how we ensure both safety and the quality of our work. Unfortunately, the commitment to each other gets lost among the organizational silos. When silos exist, staff generally don't share information or collaborate across departments. One area that's a frequent victim of this skewed arrangement is safety. Safety has historically been kept at arm's length from the rest of the organization. It's typically viewed as autonomous and separate from the work done in all other departments. Safety is, the belief goes, only applicable to frontline employees and contractors, yet for safety to work, it can't be separate. Safety needs to be the foundation upon which all other departments are built.

As we'll discuss throughout this book, safety is an attitude, not a set of rules. Service is an attitude and so is leadership. Anyone can be a manager; it's just a job title. However, in order to be a leader, you first need to have the attitude of a leader.

HOW MUCH OF A DIFFERENCE CAN A SERVICE ATTITUDE MAKE?

About twenty years ago, I was awaiting a flight on newly formed WestJet Airlines, an international carrier based out of Calgary, Canada, that started with eight planes and eighty employees. I was in a small, overcrowded boarding area when I spotted a distinguished-looking gentleman in a black suit walking through the terminal picking up empty coffee cups, discarded newspapers, and other trash. He carefully placed it all in trash containers and recycling bins.

I later learned he was Clive Beddoe, the CEO and founder of the airline. Beddoe had strong values around service and applied them to his company. Whatever WestJet could do to make the customer experience more enjoyable, he'd do it, including janitorial work if that's what was needed. To Beddoe's way of thinking, every job serves someone else. No job is less or more important than another.

When you think about it, the average WestJet customer doesn't interact with middle or upper management. They interact with frontline people like flight attendants, ticketing agents, and janitorial staff. The CEO knew that just one frontline person with a service attitude can make a huge difference. In his own unique way, he was modeling a service attitude for his staff.

Safety is an attitude that can be learned by example. Whether you're behind the counter at McDonald's or a frontline worker at a mining site near the Arctic Circle, your job ultimately serves someone else, especially when it comes to workplace safety.

SAFETY IS NOT ALL HARD HATS AND WORK BOOTS

Some may think that safety is meant for job sites where dangerous work is done. Yet for one of my client organizations, a municipal government with thirteen thousand employees, their most expensive worker's compensation claim was from an office worker, not someone wearing a hard hat.

In an attempt to reduce the cold air blowing on her desk, an employee used her rolling chair as a step stool to cover the vent with cardboard. She fell and broke nearly every bone on the left side of her body.

It's not unusual to hear from oilfield and construction companies that when they get to the job site they're impeccable in their work. Rarely is there an incident. The problem, however, is in getting *to* the job site. For instance, an employee driving a truck outfitted with heavy equipment pulls off at a truck stop, and instead of climbing down the steps, he jumps down the final two. A rollover on an ankle sends the driver home.

Safety is not just for people on construction sites. It's for drivers going to and from those sites, as well as for those who wear suits or dresses to work. For office workers, the rush-hour drive can be the most dangerous thing they do all day. Distraction and stress shorten attention spans. An ironic and dangerous example of a distracted driver is the employee who texts while speeding to make it on time to a morning safety meeting.

The problem is that most employees haven't embraced a

safety attitude. Companies say they care about safety and try to show it by spending a great deal of time and energy on creating procedures, programming, and production for safety, but when it comes to developing programs to honor their good people for their safety service, they fall short.

Companies fall short in the way they educate staff. They haven't yet realized that safety is an attitude that needs to reach throughout the company. It needs to go deeper than mere compliance with Occupational Health and Safety regulations on job sites.

According to Bill Coyne, a vice president of sales for ice cleats maker Winter Walking:

"There are three ways you can prevent winter slips and falls in a work area. 1) You can housekeep the problem with sand, salt, and shoveling, which is difficult in a large area. 2) You can eliminate access to the area. 3) You can increase traction underfoot.

"The first two options are standard safety practices in any hazardous work area. The third, increasing traction underfoot, is a voluntary option. However, using ice cleats to increase traction could be the smartest choice. Yet it's not mandatory according to Occupational Health and Safety regulations," says Coyne.

"Companies that utilize ice cleats on winter job sites aren't companies that are looking for the bare minimum. Ice cleats, in general, aren't mandatory personal protective equipment (PPE). Even in areas where there's a good amount of cold weather, ice cleats are only voluntary. So there needs to be a more proactive safety culture in the workplace."

Winter Walking has been in business for over forty years. "We've helped elementary schools, colleges, and universities," says Coyne. "We've helped companies in the oil, gas, construction, railroad, and utility industries. We've even sold to Google. It doesn't matter if your employee moves twenty feet in one direction or two miles in another. Wherever they move wearing cleats, they're going to be protected with every step they take because the protection goes with them. It's right under their feet."

Therefore, companies that want to protect their employees beyond the traditional bare-minimum occupational safety and health requirements will purchase ice cleats for their crews. An attitude geared toward safety goes beyond hard hats and steel-toed work boots to the broadest understanding of work environments—from icy job sites to office cubicles and transportation to and from the work site.

SAFETY ISN'T JUST WHAT WE DO ON THE JOB

Barbara Dunn was a flight attendant for Air Canada for thirty-two years. She's now an internationally renowned expert in passenger safety. In a 2016 airing of the CBC television show *Marketplace*, Ms. Dunn noted that most people who board an airplane are ill-prepared for an emergency.

"Our safety record is good, but it's creating complacency," says Dunn, "because people don't think something bad is going to happen when they fly. Or if something unusual does happen, they figure it won't be survivable anyway. Ask any flight attendant about the in-flight safety demonstration, and they'll tell you that most people don't pay attention. But the reality is that eighty percent of accidents are survivable."

According to Dunn, not only should passengers observe the safety demonstration, but they should also read the safety card, which includes additional information. Part of their survival strategy should be appropriate footwear. High heels are an especially bad choice for air travel. They make movement in an emergency difficult, and pointy heels can puncture the emergency slide, rendering it useless. High heels also increase the likelihood of a sprained ankle, limiting movement.

Safety isn't just about what we do at work. It's part of living in general. An attitude of safety extends beyond the workplace and past the boundaries of workplace compliance.

Unfortunately, instead of a culture of safety both in and out of the workplace, what we mostly have is a culture of mediocrity. Managers and supervisors typically care only about compliance. Like so many of us, they fail to embrace safety as a value in their own lives.

I wasn't much different myself until I got motivated and stopped smoking. I came to realize that when we start to take safety personally, the bare minimum is no longer good enough.

WHY THE RULE BOOK DOESN'T WORK

The Occupational Health and Safety Code is a rule book focused on establishing a set of bare minimums. These are the expectations of minimum safety standards with which companies and their managers, supervisors, and employees must comply. The safety code focuses primarily on process and procedure.

However, when you aim for the bare minimum, that's just what you get. When it comes to workplace safety, especially involving potentially hazardous work sites, we have to ask ourselves, "Are the minimum standards good enough?"

PENTA Building Group in Las Vegas, Nevada, thinks we need to aim higher to achieve a meaningful difference in safety. The company's operations supervisor, Steve Jones, points out a strange distinction in the Occupational Health and Safety Code. According to the regulations, if you're working twenty feet above the ground, you're required to be tied off, but if you're working over a twenty-foot open pit, you're not required to tie off. Either way, however, it's still a twenty-foot drop.

Working over open excavation is just as dangerous. Therefore, in both situations, PENTA requires their people to tie off. That's just one way that PENTA Building Group exceeds the code voluntarily and as normal practice. As a construction firm, they've made safety one of their company's foundational values. The difference this has made in their safety record is tremendous. In 2015, among their six-hundred-plus employees in Las Vegas, there were only four recorded incidents: one bruised femur, two finger cuts, and one pinched hand.

Meanwhile, other organizations of a similar size are busy chasing the bare minimum in compliance. The concern isn't for keeping their crews healthy, but for avoiding fines and court cases. However, in the long run, if what they want is to avoid lawsuits, they'd do much better to raise their safety standards and start caring about their crews.

Safety is where moral and financial principles dovetail.

Only when companies wake up to the facts will they begin to make real progress in workplace safety. It starts by embracing safety as the company culture and choosing to go beyond the minimum standards.

CREW CODE

In addition to the Occupational Health and Safety Code, every company has its own corporate safety manual. These are the processes and procedures specific to a company that meet or exceed the minimum standards of the OH&S Code. A company can make changes to a corporate safety manual, but they don't happen overnight. You have to consider the time spent retraining your people, updating manuals, and many other details. It's not the easiest thing to do.

There's also an unspoken code of "how we do things around here." This is how workers and supervisors interpret the processes and procedures in the safety manual. This unspoken code may be more stringent or less stringent than the company manual or the OH&S Code. It depends on the crews' viewpoint and the supervisor's viewpoint on safety. In some instances, it's basically a "what we can get away with" code. Unspoken codes that shortchange safety are both illegal and dangerous—the opposite of a culture of safety. However, a supervisor and crew can choose to set their bar higher and exceed the minimum safety standards.

A true safety leader with strong safety values doesn't skimp on the law or the company manual. He communicates safety expectations in an honorable way that gets buy-in from his crew. It's the only way to build respect and ensure safety on the job.

Crew members naturally want to fit in. No one ever wants to feel like an outsider on a crew. The pressure to fit in is a driving force that forges group identity. Supervisors, therefore, have a unique opportunity to establish group cohesion through a crew code that raises the bar on safety. A smart supervisor creates a work environment in which crew members will buy into safety as part of an evolving crew code. If they want to fit in and the bar is set high, they will rise to that bar to be part of the crew.

A crew needs to own what they do. Supervisors who want to build their crew code in positive ways will champion better ways to improve expectations about "how we do things around here."

MAKING THE SHIFT FROM PROCESS TO PEOPLE

In a perfect world, processes and procedures are definable and repeatable. However, when you throw in the human element, process and procedure don't always work. People are the most difficult variable to control. It is therefore vital that the human element carries more weight in safety discussions, planning, and protocols. Supervisors and safety people, however, continue to focus on enforcing process and procedure while ignoring the people involved.

How do companies begin to integrate the human factor?

The first step is to stop viewing employees as replaceable cogs in a machine. Every crew member should be valued as a person first. People know when they aren't being valued, and it reflects in their work. Minimum standards in safety breed minimum standards in morale. If the focus is on achieving minimums, staff will only be inclined to do the minimum required to avoid losing their jobs.

To value our people, companies should appreciate each member of a crew for their unique contribution to the team. A company's safety standards need to support people at the human level. Certainly, any employee can be replaced, but by any measure, moral or financial, a company is invested in each hire. They should want to keep each staff member healthy and on the job.

This is the way to win hearts and minds and build successful companies. When people feel appreciated and valued, they have reason to feel proud of their work. When they know that their employer stands by them and is willing to go the extra mile to ensure their safety, they'll go the extra mile on the job. They will take both their work and safety to heart. It becomes a matter of pride. Make them proud, and they're going to want to protect that pride.

Never will your best day, the day in which you are most proud, be a day where you shortcut safety.

The choice is either to police your people into compliance or to build a culture of safety that wins their hearts and minds. People don't like being policed. No one does, but people do like being appreciated. As a motivational tool, hitting frontline workers over the head with a rule book doesn't work. In fact, it takes much less effort to let them know they're appreciated. It's much easier to get employees to buy into a safety program when it's accompanied by real concern for individuals.

People who feel appreciated have the tendency to want to stick around. Turnover and attrition drop in company cultures of appreciation. Loyalty to the employer rises. When people have loyalty, they have ownership. When you build loyalty, you reduce turnover.

Turnover is harmful for a company and its safety program. When you lose a good high-producing employee, it takes up to a year and a half for a new hire to bring their own production level up to speed. Simply put, there's a learning curve for new employees. There are new bosses and supervisors, new co-workers, new processes and procedures. It takes time for everyone to adapt. And while adapting, a new hire is at greater risk of being hurt.

WHAT SAFETY PEOPLE AND SUPERVISORS ARE GETTING WRONG

Safety people and supervisors who lack a healthy dose of willingness to engage crew members on a human level will limit both themselves and their crews. It doesn't happen on purpose, but it does happen. The inexperienced supervisor, who doesn't know how to motivate and develop individuals on the job, ultimately has a harder time getting the job done. If there is no strategy to continuously improve employees, there's little chance of improving the organization as a whole, and that includes safety.

In fact, there are seven particular things that inexperienced and poorly trained supervisors and safety people do wrong. Let's take a look at each of these mistakes.

1. THEY FOCUS ON RULES ENFORCEMENT

When first-time or underskilled supervisors are suddenly thrust into positions of responsibility, their first priority is to not be seen as incompetent. They don't want to be the one who holds up or impacts production in a negative way. A supervisor without adequate managerial support or training tends to embrace rules, regulations, and procedures. It's all they've really got. After all, no one can challenge the rules or the law.

When it comes to safety, the bottom line for unskilled supervisors is to simply meet the requirements of the Occupational Health and Safety Code. They figure that if they stay within the law, they will keep themselves and their people out of trouble. What they strive for is minimum compliance to cover their asses. The strategies and tactics they use are centered on rules enforcement learned from their old bosses and supervisors.

A cop at the side of the highway intimidates you to slow down, even if you aren't speeding. As long as that cop is there, you will comply. However, the moment that cop is gone from your rear-view mirror, you're back to ignoring the speed limit. This is the problem when it comes to minimum standards compliance. There needs to be a safety cop on duty at all times. People know they're being watched and will comply.

In order to force compliance, unskilled supervisors and managers will employ the use of gruesome photos of dismemberment or death, gut-wrenching stories from accident survivors, threats of fines, or firing. It's hardly the way to build solid teams of safety performers. Threats do nothing to build loyalty or values-based safety cultures, but unskilled supervisors believe they are enough to scare employees into minimum compliance.

2. THEY CRITICIZE MORE THAN COMPLIMENT

Everyone hates to be criticized. Criticism is a huge demotivator. Being dumped on takes its toll on the job. Eventually the voice of the safety guy or supervisor gets tuned out. An article in the *Harvard Business Review* explains how for every negative comment made to an employee by a supervisor or manager, it takes up to six positive comments to balance it out.

Criticizing takes no skill or experience. It's lazy and really it's the least supervisors and safety people can do. Finding fault takes no expertise, and that's why it's so easy. The real work is in finding a way to turn those criticisms into positives instead.

3. THEY ARE UNFAMILIAR WITH AN "INSPECT, DON'T EXPECT" MANAGEMENT PHILOSOPHY

You've seen it happen at safety meetings. Safety people go over something once and expect that everyone got it and understood it. Later, if an incident occurs, the safety person is quick to point out that it was covered in the safety meeting. It was covered, sure, but was time taken to inspect, or make sure, that everyone understood it? In safety management, the key is to inspect for certainty. You can't simply expect; you have to follow up.

4. THEY ARE QUICK TO BLAME THEIR CREW

When "inspect, don't expect" is ignored and an incident occurs, the inexperienced supervisor is quick to blame someone else. Finding blame is what people do when they lack confidence or are insecure. It signals a lack of management skills. Blame from a supervisor can turn over staff, kill morale, and contribute to a dwindling safety culture.

After an incident, Root Cause Analysis (RCA) is used for the purpose of exposing gaps in a safety program. It's not about assigning blame, but finding the source of problems in order to build solutions and ways for improvement.

5. THEY ARE QUICK TO TAKE CREDIT

A poor supervisor is just as quick to snatch credit away from a team effort. It's like an assistant coach taking credit for a sports team's championship. It's laughable. Safety performance isn't attributable to one person. It's a shared effort when it goes right and also when it doesn't. If a supervisor or safety person wants to know what it's like to be universally despised, let them take credit away from their team just once.

6. THEY BLAME SENIOR MANAGEMENT

Openly articulating displeasure with senior management's lack of commitment to safety and resources is a knife in the heart of safety culture. Sure, it's nice to have senior management support for safety, but it isn't a requirement.

Frontline supervisors can still make and inspire excellent safety decisions without more money or a commitment from senior management. The safety person needs to build safety values regardless of senior management's level of commitment.

7. THEY FANCY THEMSELVES AS LEADERS

Anyone who thinks that their safety certification automatically qualifies them as a leader is delusional. No one follows or respects a manager or supervisor purely by virtue of their title. Leadership doesn't come with safety certification.

In fact, many times, the opposite is truer. Safety people and supervisors are viewed cautiously by the crew until they earn acceptance as leaders. A manager or supervisor isn't a leader automatically. Leadership is earned over time through many trials on the job.

COMMAND AND CONTROL IS OUTDATED

Command and control represents the tired model of last resort for poorly performing managers, supervisors, and safety people. Those who lack basic management skills default to command and control because they have nothing to fall back on but authority. It's the last resort of safety cops.

The old model of enforcement to achieve safety compliance needs to be left by the wayside. People are still being hurt on the job because a key component of effective safety management is still missing. And that component is people—highly motivated people, including supervisors who want to help grow the talents and performance of their crews and crew members who aspire to do their best work in the safest possible way.

Cops are enforcers, not motivators. Successful supervisors know how to motivate their crews to want to buy into safety. They know that safety and productivity are joined at the hip. Smart supervisors build a win-win dynamic for staff to voluntarily jump on the safety bandwagon.

The good news is that every person can find a win for themselves and their teams in safety. A skilled supervisor or safety person can be a catalyst. Through one-on-one conversations that build trust and mutual respect, employees are more inclined to hear what a supervisor has to say. They are more willing to let go of any preconceived notions that prevent them from seeing the big picture on safety. A few minutes of face-to-face conversation can build lasting trust over time.

Frontline supervisors or safety people who set an example through sincere communication earn the loyalty of their crews. Crew members become more willing to take direction and embrace the safety attitude. It's far different from command and control and much more rewarding and effective in the long term.

COMPETENCE IS MANDATORY; EXCELLENCE IS A CHOICE

Competence is the level to which you are trained. Excellence is a personal choice that goes beyond competence. Competence is expected, while excellence is voluntary.

Every company provides training to their people. Excellence, however, isn't something you're trained to. That's why two employees can get such wide disparities in results. One who works at the level of competence gets average results. The other who works at a level of excellence gets exceptional results. Both receive the same training, yet one chooses to move beyond competence. Not every employee has the same degree of motivation. It's a personal choice. Some make a conscious or even subconscious effort to go beyond what's expected.

The disparity between competence and excellence applies to supervisors as well as their crews. Some companies are perfectly happy with their command and control supervisors, as long as production keeps moving. However, it's hard to achieve excellence when supervisors or safety people set a low bar. All too many managers attempt to create a compliance culture using guilt, shame, and manipulation. They strive to achieve safety by coercion rather than by desire.

Presentations and discussions about safety shouldn't be aimed at frightening crews or turning their stomachs. Safety discussions should positively focus on safety, not negatively focus on injury. Too often safety people attempt to use failure as a motivational tactic. The message is, "Don't do what he did," rather than, "Do it this way with safety in mind." Trying to scare people into following the rules doesn't encourage them to embrace safety as a personal value. All it achieves is temporary minimum compliance, but it doesn't last. The more you do it, the more your people are desensitized to it. Next time you'll need to scare them a little bit more.

Safety shouldn't make you afraid; it should make you confident. A solid safety program utilizing effective supervisors and safety people empowers good people to make good decisions that benefit them and their co-workers.

GOOD SUPERVISOR, BAD SUPERVISOR

Let's look at a comparison: Two frontline supervisors, but each has a radically different approach to safety.

The first supervisor focuses on rules enforcement. He's the "gotcha" guy, walking around the job site looking for a reason to write you up or assign blame. It's no surprise that his crew has a high turnover. People don't want to work for him, so his crew always has new workers who aren't familiar with the safety systems. They're nervous, looking over their shoulders, not well-focused on the job at hand. This higher level of distraction increases stress and risk of incident.

The second supervisor uses a softer touch. He takes an interest in his people, chats with them regularly, knows them all well. He's taken the time to show his crew that safety is everybody's responsibility and everyone's win. His attitude and caring create stability and loyalty among his crew, so turnover is all but non-existent. The longer they work together, the more trusting they become of the supervisor and each other. In the end, they're a tight team of high performers with high expectations.

You have likely worked for both kinds of supervisors. You see this distinction in supervisors all the time, especially in the oil and gas industry. Oil and gas contractors are constantly moving from job site to job site in a very mobile industry. Contractors and subcontractors cross paths frequently and exchange information about the companies and individuals they work with. They know who the good guys are—the good supervisors who treat their people well and maintain high standards. These are the frontline supervisors who get recruited for work—and when a good supervisor is recruited to a new job, his crew usually follows.

Contractors also know which site supervisors to avoid.

They hear about incidents and safety violations, and they pay attention to how supervisors manage their crews. Poor performance gets talked about. These become the supervisors of last resort.

The bottom line for supervisors needs to be set higher. Every supervisor needs to strive to be the kind of leader that crew members will follow to the next job.

CONSEQUENCES AND CARING

Every decision and choice has both short-term and long-term results. Our overall safety strategy should instill in our people an awareness of the consequences of their decisions, both in the short term and long term.

For example, skipping a procedural step may affect safety in the short term, but what you eat, drink, or otherwise put in your body can shorten your long-term plans to keep working and earning a living.

As I described in the introduction, my whole approach to safety shifted when someone I deeply cared about told me I was needed for a long, long time. It suddenly dawned on me that my well-being really mattered, so I began to take more responsibility for it. I quit smoking and improved my diet—no more double Baconators for lunch.

The same is true in your workplace. Effective supervisors and safety people must carry a similar message to build the kind of workplace relationships that say, "I want you to be able to stick around for a long time. I want you to keep working on my team." It's the kind of relationship that builds loyalty and shared interest in safety.

Ask an employee what their vision of life looks like when they retire, and you might hear "I want to buy a motor home and travel the country" or "I want a cabin on the lake where I can fish."

Sometimes the line between what an employee is doing today and what they envision doing in retirement is disconnected. They may not be leading the kind of life or making the kinds of decisions that will get them to their retirement dream. However, their in-the-moment decisions and their visions for retirement are in fact closely connected. To reach their personal goals for retirement will take both good health and financial planning. People want to be able to enjoy their golden years, not suffer through them because of ill health.

The same planning applies to safety on the job. Nothing can put a dent in a worker's short- and long-term plans like an incapacitating injury. Poor health or injury is a recipe for early, unwanted retirement. Future plans for a life of freedom can be snatched away.

A frontline supervisor who truly cares about his crew will work toward getting each crew member to buy into the safety program and a healthy lifestyle. That's where the foundation of a strong safety culture starts. A smart supervisor leads by example, talks things out, and initiates the kinds of conversations that positively influence employees to take safety to heart and mind.

No one really wants to put their long-term security in jeopardy. No one wants to place an economic burden on their spouse or children. When supervisors move the safety discussion to a personal level, they make the kinds of connections that ensure safety for the long term.

Finally, we should ask our employees what kind of legacy they want to leave behind. In twenty years, do they want to be speaking to a safety conference about how they got hurt? Or do they want to be giving a leadership speech about how safety has made them a success? Their legacy is in their own hands. They decide by the choices they make each day.

When we consciously embrace safety as a personal value, we are more empowered to get the results we want. Good people who care about others don't roll the dice with their choices. They make conscious, safe decisions for the good of all. In the following chapters, we'll look at the practical ways to make workplace safety happen.

CHAPTER

2

THE CASE OF THE MISSING P

Most safety programs focus on the Three Ps: Procedures, Programming, and Production. These are easily replicated and adjusted to fit any function, job site, or place of employment. However, one P is missing from this list. Let's look at the first Three Ps on our way to finding the missing fourth P, which is most important in the new safety model.

1. PROCEDURES

Procedures and processes must be followed in order to stay within safety law. The Occupational Health and Safety Code specifies the practical how-tos: everything that must be done on the job to ensure safety. If there is an incident, the first thing to look at is whether workers followed the proper procedures.

2. PROGRAMMING

The second P is programming, the training and repetition of training. This is where we drill information on procedures into our people over and over again, whether it's at a weekly safety meeting or the morning tailgate meeting. During these meetings, your people are programmed with lots of information and guidelines. Then this information is reinforced in bite-sized chunks during crew meetings. The idea behind this is that the more often your people hear this information, the greater effect it will have on their work.

3. PRODUCTION

Production is how the job gets done after the programmed training of your people in the proper procedures. When enough focus has been placed on the first two Ps, the last one, production, should follow safely. That's the ideal, but it doesn't always work out that way. The reason lies in the missing P.

4. THE MISSING P

The problem with the Three-P model is that it works best with robots. A procedure is created, it's programmed, and the industrial robot starts to produce. However, when people are brought into the equation with their human emotions and personalities, things can get a little more complicated.

The logical method and approach of the Three Ps misses the human factor. In our work-a-day world, the Three Ps just aren't always effective. As it turns out, the missing human element in safety is the most crucial part.

The fourth P is people.

So much time, effort, and money is wasted on enforcement tactics because so little time is spent on encouraging employees to buy into safety. The energy spent on enforcement could be better spent building teamwork, morale, and camaraderie.

As it plays out in the real world, a team that has adopted safety as a personal value is better equipped to make the kinds of decisions on the job to ensure safety. Instead of mere compliance with procedures, programming, and production, a safety-oriented crew draws from a deep well of mutual caring and connection. When crews themselves become safety leaders, the need for safety cops disappears. What you have instead is a team of solid safety leaders who perform and produce at a higher level.

Crew members with names, families, and lives outside

of work are the ones who are affected when there's an incident. Therefore, when we discuss safety, what's most essential is its human face. This is the level at which supervisors need to connect with their crews.

THE M4 METHOD

The new safety model picks up where ordinary safety programs leave off.

I developed what I call the M4 Method as a way of integrating people into the strategy of building solid safety programs. The M4 Method doesn't entirely eliminate the Four Ps, but rather builds upon them. It is a tool that's used to directly help supervisors and safety people build on-the-job relationships that support safety from the ground up.

The M4 Method marries four critical components to achieve the next level in safety:

- Management
- Meetings
- Marketing
- Motivation

M4 is not designed to replace your existing safety program. It is designed to augment your safety program by focusing specifically on the fourth P: People.

Supervisors manage people and make decisions that are in alignment with the Occupational Health and Safety Code and company safety manuals. Meetings are conducted to inform and seek input from people. Marketing is aimed at communicating with and rallying people to the common cause of safety. Motivation is entirely people centered—machines and programs have no motivation, only people do.

Let's take a look at the four components of the M4 Method.

1. MANAGEMENT

The first component of the M4 Method is management, an essential skill that inexperienced supervisors and safety people tend to lack. So many people in supervisory and safety positions ascend to their jobs without rudimentary management and people skills. When this happens, everyone suffers. Supervisory staff who are poorly equipped to manage their crews wind up flailing about with nothing to fall back on but the rules. They struggle to assert authority because they miss the key component: people skills.

When people skills are present, the results can be astounding. Good supervisors and safety people understand how to create a support system that builds trust in their crews. They know how to motivate people on the job. Good supervisors and safety people know how to get frontline employees to buy into safety.

2. MEETINGS

The second M is meetings, which are legally required by the Occupational Health and Safety Code. In fact, the only department in an organization legally required to hold meetings on a regular basis is the safety department.

Given the legal requirement to meet regularly, you'd think that safety meetings would help establish significant connections among safety people, supervisors, and crews. Yet so much of that potential is often squandered. Safety meetings could be used to inspire employees, create dialogue, build trust and camaraderie. They could be used to help remove the mental barriers to embracing safety as a personal value. However, more often than not, safety meetings are used to review paperwork and inspections, rules and procedures. They become boring and mind-numbing for employees who wind up questioning the point of meetings at all. And who can blame them?

Nowhere in the Occupational Health and Safety Code does it state that meetings are required to be boring. The code certainly doesn't require safety people and supervisors to regurgitate PowerPoint presentations of endless streams of bullet points or to read slides verbatim from the front of the room. Meetings shouldn't be corporate karaoke. If you're showing PowerPoint slides of paperwork, admit it, you're out of ideas.

Instead, meetings should be opportunities to strengthen and connect with teams. Each meeting is a chance to change the "groupthink" on safety. When people are focused and engaged in the shared responsibility of their own work, they become energized. They are primed like a sports team to get the job done efficiently and effectively without injury.

3. MARKETING

The third component of the M4 Method is marketing. We don't usually think of marketing as having anything to do with safety, but they are connected. Marketing is all about communication, and communication involves getting a message across. When it comes to workplace safety, the message needs to carry across all groups of people. To do that effectively requires a solid marketing strategy.

Supervisors need to know their crews. They need to understand their crew's likes and dislikes, what motivates them and what turns them off. Just like marketers who sell products, safety people and supervisors need to sell the idea of safety. The crews are the target market.

Every organization, therefore, needs to create an internal communication strategy to reach staff. The company's commitment to safety goes a long way in winning people over. Marketing of safety needs to inspire people to want to be part of it. The messaging needs to be so clear that employees finally come to the realization that, "I get it. I can buy into that."

4. MOTIVATION

The fourth and most crucial component in the M4 Method is motivation. Without motivation, nothing works. Motivation is what gets everything else done. Motivation is what makes people stay on task and focused on workplace performance. It's what holds the M4 Method and the Four Ps together. Motivation is at the heart of safety culture.

Motivation is of foremost importance to the frontline supervisor. Supervisors who aren't motivated to reach their crews at the human level will revert to policing and rules enforcement. And if they themselves aren't motivated, you can be sure their crews won't be either. What results is a stagnant safety culture. Eventually, production will suffer as well.

When senior management isn't motivated about workplace safety, there's high turnover in the ranks. Word gets out that safety is not taken seriously, and high performers look to other companies for employment. Low morale and low motivation go hand in hand. People know that the best place to work is also the safest place to work. A robust safety culture provides a competitive advantage to attract higher-performing employees.

People are motivated when they have goals and achievements in mind. Companies should want to be the best at their game and set industry standards. Too many organizations, however, measure their safety standards against industry averages. No company ever achieves top results with average performance. In today's competitive marketplace, middle-of-the-pack performance goes nowhere.

Can you imagine a sports team approaching the new season with a mission to "just be average this year?" Average teams don't win championships. They get eliminated from championship contention early. Championship teams are highly motivated and value their players. In the workplace, as on the field, motivation and teamwork bring home the gold.

NOT JUST PLAYING IT SAFE

As we make the shift from process to people, our whole way of thinking about safety changes. We move beyond the rule book and minimum standards to higher expectations. We go from compliance to community. We go from playing the game of enforcement to coaching crews in best practices.

You'll be surprised to find that instead of grudging obedience, you'll get enthusiastic team players. You'll no longer just be playing it safe, but setting a gold standard in safety for yourself and your crews.

So let's set aside the rule book and take safety to the next level. In the following pages, we'll take a closer look at the four components of the M4 Method and how they take up where the rule book leaves off.

CHAPTER

3

MANAGEMENT: RELATIONSHIPS MATTER

Effective management, the first component of the M4 Method, requires effective relationships. Unfortunately, potentially positive workplace safety relationships can be undermined by safety cops. These are safety people who police a work site with the rule book tucked under their arms, looking for opportunities to wield their enforcement power.

Safety cops focus on rules and regulations to achieve compliance. Generally, they get what they want through fear and guilt tactics, sometimes resorting to intimidation. They aren't there to build relationships. In their view, relationships would compromise their effectiveness. After all, no one wants to report a friend or to be seen as playing favorites. So safety cops choose an adversarial role instead.

They are big on authority and believe their position entitles them to respect. They expect compliance in the same way a cop at the side of the road expects you to slow down out of fear of being stopped.

This isn't a good model for workplace safety management. As we saw in Chapter 1, good supervisors and safety people are leaders who embrace a strategy of coaching and mentoring crews for better performance. They inspire their teams by rolling up their sleeves and engaging one-on-one. They educate their crews by demonstrating practical applications of processes and procedures. It's an approach that builds confidence and mutual support.

Jim Lundrigan models this approach. Jim is operations superintendent at Nickel Rim South Mine in Sudbury, Canada. He takes a hands-on approach to workplace safety. He gets out of his office and into the field three mornings a week to talk with crews.

"I have the whole mine and a surface plant to oversee, so I go underground and to the surface plant where I can engage individuals," he says. "I became a supervisor way back when, but I'm not an engineer or a tradesman. So by talking to the guys, I'm actually learning. There are a few specific things I get out of doing these one-on-one site visits.

"After twenty-seven years, I've learned bits and pieces about every trade. I can now maintain a conversation in any planning meeting or any type of risk review meeting because of what I've picked up from talking to my people. By asking questions and learning, I've found weaknesses in our own standard operating procedures.

"The guys I talk with are so experienced; they see it for themselves right away. I may ask an innocent question like, 'Why do we do it this way?' and the reply is, 'Well, I don't know.' So I ask if there's a better way to do it, and usually they can offer a better way.

"You wouldn't believe the safety initiatives that have come from the crews after some of those interactions," Lundrigan says. "They've got the knowledge and expertise. It's just that sometimes they need a little push to speak up. Just by asking these questions, we've really developed a lot of innovation. Contractors like working with us because of our reputation on safety."

Lundrigan's approach to safety is a solid example of what

hands-on learning, mentoring, and coaching can do to improve safety management and supervision. If you truly want the best from your crews, you will have to get out of the office and into the field. You must engage your crews in dialogue, ask questions, and involve them in helping build a better safety program.

IN EVERY JOB, ONE RULE RINGS TRUE

My first paid job was as a twelve-year-old salesman on a Dickie Dee ice cream bicycle. I worked on commission with no hourly wage. In 1973, popsicles cost a nickel and ice cream drumsticks were a quarter. The ice cream bike was a single speed that weighed six hundred pounds fully loaded, and my route was a hilly, blue-collar town in Renfrew, Ontario.

Like most families in Renfrew, mine didn't have much money. My dad was an office manager at a tire shop, and my mom was an elementary school teacher. I was working to save up for a ten-speed bike that cost $125. If I wanted it, I had to earn it myself.

I quickly learned what time workers at the local factories took their breaks. I rotated days between the Coca-Cola bottling plant, the Playtex factory, an iron foundry, a furniture factory, and others. Most of the factories had no air conditioning, so a frozen treat always hit the spot. I noticed the camaraderie those factory workers shared. It wasn't uncommon for one guy to step up and say, "We've got five guys here, so it's five cones on me." It seemed to me that whatever they were doing, they were in it together.

The following year, I worked part time in a pro shop at a golf course. That's where I learned about customer service and doing a quality job. My boss was a golf pro and an intense manager who was insistent on routines, procedures, and presentation. Everything had to be done just so, no surprises. The better you did, the more you moved up.

Later on, I worked at our small town's first radio station, which my father and a couple of local business owners started. I emptied trash cans and helped out wherever I could. I learned that the best announcers were the ones who connected with their audiences on a personal level. It started me on an eighteen-year career in broadcasting doing jobs from sales rep to on-air announcer to supervisory and management positions.

Through all of those early jobs, I found that one thing trumped everything else—and that one thing was relationships. How I related to people would make or break my sales success. It determined whether or not I was kept on the job, how much I made, and whether or not I was hired back the following summer.

When I entered the work world as a young adult, I found that the lessons I'd learned about relationships at work still held true. Experience showed me that if you're not good at relationships, you don't earn respect at work and you don't go far in the world of work. The best managers and supervisors I've met on the job all had one thing in common: they were good at relationships.

SAFETY AS A PERSONAL VALUE

After giving a safety leadership presentation at an electrical generating station, I was approached by one of the employees. He wanted to ask my opinion about a recent company blanket policy that required employees to wear hard hats in the yard during smoke breaks. The fact is that a 100 percent policy on safety everywhere on-site removes ambiguity. When employees understand that safety policies aren't punitive, but meant for their own protection, they are more likely to embrace safety as a personal value. Instead of looking for excuses to take off their safety gear while still on-site, they'll be looking for ways to be more proactive about their own and others' safety.

The supervisor who enforces safety rules all day long and then goes home to cut the lawn wearing flip-flops hasn't taken safety on as a personal value. There's a big difference between tolerating safety rules and taking them to heart. You can't set a good example about safety if you don't believe it yourself.

Workers on the job can easily see through a supervisor's veneer of pretense. Supervisors who are all about rules for rules' sake don't earn the kind of trust and respect needed to be safety leaders. However, those who model safety on and off the job are the ones others will follow instinctively. They become the safety leaders who inspire others to buy into safety as a personal value. Attention to safety can actually build tighter and closer relationships on the job.

Once safety people and supervisors learn to spend less time on enforcement and policing and more on building safety as a personal value, they can focus on improving productivity and production. As safety builds stronger relationships on the job, it also improves the bottom line.

THE FRONTLINE TEAM

In a recent discussion on a LinkedIn bulletin board, a safety person claimed that their purpose was to protect their people from unsafe acts. I'm sure the person meant well, but the comment was an unfortunate choice of words.

Safety people are not supposed to "protect" workers from

unsafe acts. That's the job of a doting mother who doesn't trust her kids to wear mittens when they go sledding. As a safety leader, you're working with adults who have the ability to reason and make their own decisions. The job of a safety leader isn't to stand over them and make decisions for them. The real work of safety people and supervisors is to educate and inspire employees to make excellent decisions about how they approach their work, mitigate their risks, and make good choices.

Like a lawyer or accountant who gives advice when asked, safety people should be considered experts who understand best practices and can guide and mentor. A good safety person understands that they are involved in a collaboration with employees. It's a team approach.

When we talk down to employees as if we were their mothers, we create a disconnect. That's not what we want. The goal of good supervisors and safety people is to empower their crews. We need to get the idea of being a protector out of our heads. We're not superheroes. We're guides to best practices.

When one of the partners in a relationship talks down to the other, the relationship is doomed to fail. People need to feel valued. They need to feel that they matter. They need to feel important to the relationship, whether it's a marriage or a work relationship. As a supervisor, if you can't make your people feel they really matter, they won't stick around.

A good supervisor strives to let each crew member know they are valued as an individual. You don't have to like them personally to let them know they matter and that they're an essential part of the team. If you can't find anything about them that benefits the team, either they are wrong for the team or you are. Grudges and personality conflicts make for increased risk. As a supervisor and safety person, you have to work diligently to build trust between yourself and crew members.

Supervisors should be getting to know their people at a personal level and leveraging those relationships. It's okay to have better relationships with some crew members than others. People have different personalities, strong points, and weaknesses. When you build strong, trusting relationships, you can actually leverage your best people to become your ambassadors. They can encourage others to buy into safety. This helps break down the "us versus them" culture. Once individuals embrace the values of safety and relationship, teamwork and mutual caring become the status quo.

WHY ENGAGEMENT IS KEY ON THE FRONT LINE

The polling firm Gallup releases findings on the disengagement levels of the North American workforce. Consistently for nearly a decade, the percentage of the workforce not actively engaged in their work has hovered around 70 percent.

It's not hard to understand how having only three in ten employees engaged in their work could create problems in safety performance. Simply put, if there's no connection to the work, there's no connection to safely doing the work.

If you're a supervisor, you can't realistically expect an HR department to solve your employee engagement problem for you. Supervisors have to work harder to create better employee engagement on the front line. When an employee's mind isn't on the work, there's a greater risk of safety incidents. That's when it falls on the supervisor to improve the engagement levels of the team.

As we have already seen, frontline supervisors are given very little support with regard to management skills. Without basic skills in management, the task of engaging employees becomes infinitely more difficult to accomplish—not impossible, just more difficult. Yet with only 30 percent of employees engaged in their work, resources are still not earmarked for adequate management training.

Rarely is there discussion of how incident levels rise in direct proportion to employee disengagement. If the argument can be made that distraction behind the wheel impairs one's ability to operate a vehicle safely, it's not a stretch to see how disengaged workers compromise safety.

On the front line, it's left to supervisors and safety people to handle employee disengagement themselves. It's where the buck stops. On the front line, employee engagement is the supervisor's responsibility.

FRONTLINE MANAGERS MAKE GOOD PEOPLE STAY

When people feel valued for the work they do, they engage better in their work and are more willing to find ways to contribute. When engagement is high, the focus on safety increases. When employees feel that what they do matters, their loyalty also increases. Their loyalty may not be to the company, but rather to the supervisor who values their work and communicates genuine appreciation for their contribution to the team.

People are far more willing to stay with a supervisor when they feel appreciated. Moving to a new job under a new supervisor is always risky. It's untested ground. A valued crew member is more likely to stay put. In an environment of trust and respect, employees don't go seeking greener pastures. As the first line of contact between the frontline employee and upper management, supervisors play a key role in retention.

Even in a company with a lousy corporate culture, a good supervisor can maintain high levels of loyalty and low turnover and attrition. Conversely, even in a good corporate culture, an unskilled supervisor can be the reason for high turnover and high attrition rates.

Safety departments are quite adept at tracking lagging indicators when it comes to safety performance. But one lagging indicator that they may not be tracking is the department or crew with the highest incident rates and the highest staff turnover rates. It would be no real surprise to learn that a supervisor with the highest turnover rate will likely also have the highest number of incidents. Just as a malfunctioning machine part may need to be swapped out or retooled, that supervisor needs to be either replaced or retrained.

High turnover rates create a constant stream of new employees on a job site. New employees are not up to speed their first day on the job. It takes up to eighteen months for new staff to be performing and producing at the same level as a veteran crew. That can have a dramatic effect on safety performance and production.

YOU SPEND MORE TIME WITH YOUR SUPERVISOR THAN YOUR SPOUSE

The relationship between an employee and supervisor matters greatly to both safety and production. An employee actually spends more waking hours each day with their supervisor than with their own spouse. Supervisors, therefore, need to ensure the quality of that relationship.

How can a supervisor build and maintain the employee relationship? Three main things are needed:

1. PEOPLE SKILLS

Without good people skills, you won't have good relationships and you won't be a good supervisor.

2. RELEVANT KNOWLEDGE

This is knowledge that's relevant to the work your company does. You don't have to understand how to do the job as well as the guy who actually does it, but you should have rudimentary knowledge of it.

3. PERSONAL DESIRE

This is the desire to perform at the highest level in your own position.

Any two of the three is not enough. You can have people skills and relevant knowledge, but if you don't have personal desire, you're going to fall short. Knowledge and desire without skills will fall short. Skills and desire without knowledge will fall short.

If you've got all three, you have the foundation upon which to build solid workplace relationships. So you've got to keep making sure your skills stay sharp.

PROJECT OXYGEN

Google prides itself on staying sharp. The company is highly respected for its technical expertise and drive to continually improve. Google has thirty-seven thousand employees, including five thousand managers, one thousand directors, and one hundred vice presidents. So when Gallup releases disengagement numbers of 70 percent, even Google pays attention.

In 2009, Google launched their engagement initiative called Project Oxygen. They wanted to get a handle on reducing unnecessary staff turnover. Google also wanted to ensure that their employees were satisfied in their jobs. They wanted to know if the relationship between employees and supervisory staff was working in a positive way. Project Oxygen consisted of a series of employee surveys asking staff what they believed were the attributes of a good supervisor. Here are the top-eight findings:

1. A good supervisor is a good coach.
2. A good supervisor empowers the team and doesn't micromanage.
3. A good supervisor expresses interest and concern for team members' success and well-being.
4. A good supervisor is productive, results oriented, and always improving.
5. A good supervisor is a good communicator with the ability to listen and share information.
6. A good supervisor can help employees in their career development.
7. A good supervisor has a clear vision and strategy for the team.
8. A good supervisor has key technical skills to help or advise the team.

You'll notice that the first seven items on the list have absolutely nothing to do with technical expertise. Yet technical expertise has always been a signature trait of Google managers. Instead, employees felt there were seven skills that were even more important. Google staff wanted their supervisors and managers to possess people skills. They placed higher value on their ability to create and maintain good relationships.

The takeaway for every company, whether in construction, resources, or manufacturing, is that the attributes employees universally want most in supervisors are strong people skills.

HOW CAN YOU BECOME MORE EFFECTIVE AS A SUPERVISOR?

Let's take a look at a few practical things you can do to improve the quality of your workplace relationships.

Whether you're a supervisor or a safety person, relationships matter, so you'd better get good at them. You can't fake your way to solid business relationships. However, you can apply yourself and learn how to build workplace relationships that truly connect.

LET GO OF THE NEED TO BE RIGHT

The first, and easiest, thing you can do to become more effective as a supervisor is to let go of the need to always be right. When you have a need to be right, you're too busy insisting someone else is wrong. For employees, that's a tough environment to work in. Employees who are made to feel they are always wrong have no loyalty to the job or the supervisor, and no engagement at work.

So let go of your need to be right. Find ways to make your crew members right. Find ways to elevate their ideas and suggestions. Give them the opportunity to shine. Demonstrate to them that you value their ideas and their contributions, and they will give you respect and trust you as an adviser. This will allow you to focus instead on mentoring and coaching your people. You become an influencer instead of an enforcer. And when you've become a welcomed influencer, you can inspire your good people to do the right things, instead of pointing out the wrong things they are doing.

BE YOUR AUTHENTIC SELF

Once you've let go of your need to be right, the next area of focus is to concentrate on being the same person on and off the job. You're not playing the part of a supervisor or safety person in a community theater production. It's not a series of lines you learn. You have to bring your authentic self to the job.

If you care about people, you can care about your employees too. If you're capable of selflessness at home or in your community, you can be selfless on the job as well. If you can have empathy when your child skins their knee, you can have empathy for a crew member or employee who is going through a difficult time. If you can comfort and care for people outside of work, you can do the same for people at work.

The litmus test of your authenticity is to ask yourself whether the way you conduct yourself at work would impress or scare your own child. Could you bring your child to work and not have to change your behavior or interactions with your crew? If you can be the same person at work as you are when you're with your loved ones, you'll be a more effective supervisor.

BE CAREER MINDED

Start treating your supervisory or safety position as a career and not just a job. Consider the example of professional athletes. Pros don't show up to a game hoping to "wing it" to victory. They spend countless hours practicing and fine-tuning with performance coaches and sport psychologists. They incorporate new ideas and strategies to up their game. The best athletes are the highest paid because they meet the highest expectations.

What are you doing to improve your game? If you want to be an exceptional supervisor or safety person, you need to hone your craft. You need to schedule time on your daily calendar for professional development. If you don't up your game, neither will your crew. Work is performed better by people who improve their skills. Crews respond to supervisors who set higher standards, and when the quality of the work gets better, so does safety.

When you boil it down, the bulk of your work involves communication in one form or another. To become an effective supervisor—a professional supervisor at the top of your game—focus on developing effective communication skills. Invest in yourself and your career. Read books on workplace management, attend seminars, watch videos. Make the extra effort and so will your crew. It's time to enter the pro ranks of your industry.

DON'T SUCK UP THE OXYGEN

There's no need to prove you're in charge. You don't have to dominate. You have to listen and respond. Communication doesn't work without careful listening. Hear people out before formulating a response. Communication isn't a contest or an assertion of authority.

If we as supervisors or safety people suck up all the oxygen, there's none left for the team. Encourage your crews to ask questions, comment, and offer ideas and suggestions. Create an environment that allows them to breathe life into the team. When you ask questions and listen to your crews, you'll actually find better ways to get the job done.

This is particularly useful in toolbox or tailgate safety meetings. When safety people give long-winded, forty-five-minute speeches that cover twenty-seven different topics and seventy-two slides, all you're doing is sucking up air. Your people find the information dump stifling and don't really benefit from it.

So change your approach. Coaches and mentors don't take center stage. They want their people to have enough air to flourish. The best coaches in sports establish strong relationships with their players. Players are the stars. Coaches are in the background fine-tuning the talent on the field. As an effective supervisor or safety person, your best strength is your influence.

The best coaches take the long view and make incremental changes one element at a time. They don't give a player thirty-seven things to work on at once. They concentrate on short, actionable pieces of information that can provide direct results. As a frontline supervisor, concentrate on your crews making small, day-to-day refinements. Let them know you're in their corner and invested in their game, not yours.

BE ENGAGED

Employees engage in their work in direct proportion to

their supervisor's engagement. The supervisor or safety person who is not actively engaged in his work has little chance of inspiring an engaged crew.

This means you need to be actively involved in safety and communicate a genuine passion for it. Without real engagement, all you're doing is trying to get employees to follow the rules. High levels of safety and production on the job require meaningful buy-in by employees. Genuine passion for safety creates trust and respectful relationships.

Relationships of trust are the cornerstones of safety. To get there, you'll need to be an engaged supervisor.

FORGING SAFETY LEADERS

Strong management skills coupled with strong work-place relationships is a winning combination. The two are inseparable in the service of safety. In the M4 Method, management comes first, because it's the crucible where leaders are forged.

Whether or not you have the disposition, temperament, and mettle of a leader—and not just a manager—is entirely up to you. In the next chapter, you'll see how the way you view people can make or break your success on the job.

SUPERVISOR PRIORITIES FOR SAFETY

• • • • • •

- Focus on your crew.

- Build a crew culture.

- Improve trust and respect.

CHAPTER

4

MANAGEMENT: PEOPLE VIEW

"The way you see people is the way you treat them, and the way you treat them is what they become." Wise words from philosopher and writer Johann Wolfgang Von Goethe.

Supervisors and safety people who use their titles to boost their egos don't usually have a favorable view of people, especially those who work under them. They don't see their people as equals or value their contributions. They tend to think that their title or position carries some special authority and prestige.

You can use your feelings toward people, your perceptions and preconceived notions about their skills and personalities, either as ammunition to assert your alleged superiority or as tools to help get the best performance and engagement from your team.

As a supervisor or a safety person, you can either elevate or drag your people down. You can use your influence to improve performance or you can follow your people around waiting for an opportunity to scold them. How you conduct yourself on the job site will vary greatly based on how you view your position and how you view the people you work with. We call this "people view."

When you adopt a positive people view, it will motivate you to acquire managerial and coaching skills that can positively influence your people to better performance. With a positive people view, you can use your skills for good. Alternatively, you can use your position to assert your authority and make your people feel small and insignificant—but that would be a real waste of everyone's time and effort. You'd be using your position for the wrong reasons.

In this chapter, we'll be working together to build a foundation on which you can build a positive people view and outlook.

THE PROBLEM OF LOW EXPECTATIONS

A people-oriented viewpoint is an important component to building solid performance in safety. Sadly, unskilled supervisors and safety people traditionally don't have much faith in their employees to make good decisions. They have low expectations.

Many supervisors and safety people bring the example of their past bosses to work with them. They use fear and intimidation, and assert their authority because those were the strategies their past managers employed. There's no question that your management style will largely be based on the management styles that you've been exposed to in the past.

Some supervisory people believe that they have to "act" this way because they don't believe that their crews and employees are capable of much more. They "crack the whip" because they may view their own people as "grunts," "cowboys," "idiots," and even "mouth breathers." I've laid witness to many a safety person and supervisor who believes that their rough-around-the-edges crews are worthy of these descriptors. Imagine yourself working for a supervisor who viewed people this way. Imagine the level of resentment and mistrust the crew would feel.

When you lack a positive people view or have a low level

of respect for your employees, you can't hide it, at least not for long. If you believe that you're working with grunts, your language, behavior, and expectations will reflect it. When you have low expectations of their abilities, you dumb down your language and share just enough information to get bare-minimum compliance. You don't try to raise the bar of discourse or expectation. Instead, you find yourself chasing after employees just to get the bare minimum done. Your position evolves into one of a policeman or enforcer. That's an impossible place from which to lead a team of winners.

Instead of viewing your people as grunts, if you were able to view each one as a competent individual capable of making good decisions, your expectations of their abilities would be heightened. You wouldn't just enforce the rule book. You'd set a higher bar, and your crew would reach for it because that is the expectation.

It isn't easy. If you feel that you need to be extra vigilant or suspicious with some crew members, it's tough to view them in a positive way. If you believe that someone can't be trusted, you're not going to trust them. Instead, learn to build trust by giving crew members opportunities to prove your suspicions wrong.

HOW THE MARINES VIEW PEOPLE

Armand Audette spent time in the United States Marine Corps. Since re-entering civilian life, he has taken much of what he learned in the corps and transferred that experience to the workplace. He's just beginning his journey into the world of becoming a safety person.

The transition from military service to civilian life has been a major readjustment. "One of the things missing in my life since I got out of the corps," says Audette, "was having a team to care about and do my best to look out for them. Where I finally found that team in the civilian world is in safety."

Audette has found similarities between the teamwork and culture of safety among civilian work crews and military units. "It's critical to be tuned in and looking out for each other," he says. "It kind of frees you up a little bit because everybody else is doing the same thing."

He points out that teamwork really works when everyone is invested at an emotional level. "You absolutely have to care about the people in your team. You might not like some of the things they do, or their personalities might rub you the wrong way, but you have to care about them. If you don't, it wrecks the cohesiveness of the team. You can have a mediocre team, or you can have a team that embodies esprit de corps [group spirit]. When you have a team that eats, sleeps, and breathes esprit de corps, you can ask them to do anything."

Audette's military experience taught him that in order to be a great leader, you first have to be a great follower. "One of the main things in the corps is that every soldier knows the next person's job. It's almost built-in mentoring. Under optimal conditions, you have four men in your team, but most times are not optimal, and you end up with three-man teams. Each of the three covers the fourth man's responsibilities. It prepared me well to be in the civilian world and to be able to adapt. It helped me to become a team leader and to recognize what it takes to get there."

When it comes to how true leaders view, treat, and mentor people, Audette says, "If you want people to aspire to something better, treat them as though you already see them there. They're capable of being more."

DO PEOPLE WORK UNDER YOU OR WITH YOU?

From the employee's perspective, there's a great deal of reward in working with (not just for) a supervisor or safety person who views the crew in positive ways.

If you view people as working under you, then you tend

to think of yourself as superior. If you embrace the attitude of working with and alongside them, you become a people-oriented leader. True safety leaders don't think of themselves as superior to their crews. A leader views his people as equals who can one day do the leader's own job and maybe do it better.

There's a difference between being in charge and being a leader. When you're in charge, you may have more paperwork to do, but it doesn't mean you're a leader. You've just taken on more responsibility. A true leader uses his position to advocate for his team and strives to build leaders among them.

As a supervisor or safety person, consider whether or not you're working with the goal of improving your people or whether you're working to keep them following rules. If you're policing, you're not building safety partners. When your only focus is on catching mistakes, you're not focused on teaching them how to do things right.

A positive people view encourages people to want to do the right thing. A supervisor who works with his crew will focus on their potential. He or she looks for teaching moments, instead of relying on manipulation or punishment.

A people view involves a willingness to advocate and stand up for your crew. It involves championing their abilities and considering their ideas. A positive people view is the viewpoint of a leader who builds morale while increasing safety performance.

BUILDING RELATIONSHIPS BUILDS INFLUENCE

Marty Park is a serial entrepreneur and business coach in Calgary, Canada. Since the age of nineteen, Marty has owned and operated a multitude of businesses, from restaurants and learning institutions to the work he currently does as a business consultant for entrepreneurial firms. Along the way, he's learned a thing or two about creating relationships at work. He bases his business success in part on having a positive people view.

Marty's belief that most people are just good people who need good information to make good decisions puts him in a position where he becomes a solid center of influence with his clients. They listen to what he has to say because of how he elevates their worth in his own mind. Marty is keenly aware that the way you view people determines your influence in relationships and business outcomes.

For some supervisors, the toughest part of developing a positive people view is learning to set aside bad feelings for certain individuals and focusing instead on their value to the team. What supervisors tend to find out is that as their viewpoint changes, so does the quality of the work they get. Good feeds on good.

You can get to a point where you begin to see that everyone you work with is essentially good and wants to take pride in their work. The crew then becomes an exceptional collection of individuals, each bringing a unique set of strengths to the job. Even the slow adopters eventually buy-in when they see the rest of the crew achieving at a higher level and holding higher expectations of their teammates.

THE D STUDENT

The D student struggles to get through school without failing. He is always just one bad test away from turning that D into an F. The enforcement of safety rules as a minimum standard is the corporate equivalent of the D student—just one bad test away from a failing grade. When you focus only on meeting minimum standards for safety, you are like the D student squeaking by on a minimum passing grade.

Supervisors who only strive for minimum compliance wind up shortchanging productivity. When increased production is called for, they are willing to shortcut safety to get there. Your most productive day should never be unsafe or illegal. Your best work—the kind you can be proud of—can only be reached on a foundation of safety. Pride in accomplishment comes only from knowing you did everything right, paid attention to the law, and used your best management skills to get results.

A NEW MARKET-DRIVEN STANDARD

In the late 1990s in the province of Alberta, a program called the Certificate of Recognition or "COR" was born. Basically, it's a prequalification requirement for contractors working on both public- and private-sector projects. COR is aimed at driving best workplace behavior and practices. To earn it, an employer has to demonstrate that quality health and safety management systems have been developed, implemented, and evaluated on an annual basis through internal and external audits.

During these safety audits, an external auditor will come in and study an organization's safety program, looking for gaps and inconsistencies. They also make sure that the safety program is actually being implemented the way the organization claims it is. Once bestowed, the Certificate of Recognition is good for three years, provided the organization performs annual maintenance audits and continues to comply with the terms and conditions of the program.

Since its inception, the COR program has spread and become popular throughout Canada. It is a market-driven initiative that far surpasses the minimum standards of the Occupational Health and Safety Code. COR has, in fact, become a minimum requirement for bidding on contract work in many industries. It's the new market-driven standard. Companies without a Certificate of Recognition are finding themselves on the outside looking in when it comes to securing work as preferred vendors and contractors for big industry projects. This puts tremendous pressure on companies that are used to complying only with bare-minimum standards.

As the market has raised its requirements for participating on the big jobs, frontline supervisors are feeling the pressure. They are coming to realize that they'd better start making friends with safety, or they'll be left behind. Safety has become increasingly integrated with every aspect of how an organization does business. Safety is the present and future of how business is and will be done. As a frontline supervisor, now's the time to buy in, or you'll be fighting for a piece of the scraps the big players leave behind.

LIVE IT, DON'T FAKE IT

When it comes to a positive people view, don't just give it lip service—believe it. People can tell when you're being disingenuous. Supervisors who care about safety don't let anything get in the way of preventing their people from getting hurt.

If you care deeply enough, you'll work to give your people the skills they need to protect themselves from harm. As a result, they will respond in a positive way. People will care about their safety when they know that you care about them. That's how it works. So live it, don't fake it.

FROM SLOGAN TO FRONTLINE REALITY

"We don't hurt people" has become the new rallying cry for companies. However, well-intentioned words and mottos aren't enough. We live at a time when there are more safety professionals and better procedures than ever, but a full-hearted embrace of safety is still lagging. Incidents are still on the rise, and it makes you wonder how this could be happening.

The fact is that we haven't reached a point of saying, "Enough is enough." It's no longer enough to just track numbers and congratulate ourselves when the numbers are low. We can't be satisfied with measuring our safety performance against the industry average. We can't settle for average. We need to get every safety person, supervisor, and manager to buy into a positive people view. We need to force our organizations and ourselves to rethink safety.

There is no way that companies can attain exceptional safety performance when only a handful of employees adopt safety as a personal, non-negotiable value. The whole organization needs to be behind it. The entire organization needs to want it so badly that they are willing to challenge convention in everything they do. When every single person in an organization adopts a people-centered viewpoint, when they bring a positive people view to the people they work with, workplace safety can shift from a slogan to a frontline reality.

PERSONAL ASSESSMENT

· · · · · ·

What's more important to you as a supervisor: Corporate safety performance or how you're viewed? Which motivates you more?

CHAPTER

5

MEETINGS: SPINNING THE REQUIREMENTS

After management, the second component of the M4 Method is meetings. Safety meetings are a legal requirement. In fact, they're the only meetings that are legally required besides the general shareholder's meeting. By law, inspections and incident reviews must be covered in safety meetings, and so the lion's share of meeting time is focused on paperwork.

Of course, paperwork is important, however, there is far too much focus on forms, reports, and inspections at safety meetings. When you are showing PowerPoint slides of paperwork, you're not adequately focused on the opportunities safety meetings can offer. You're focused on compliance instead of using safety meetings to rally your people to teamwork, performance, and safety culture.

Although safety meetings are a legal requirement, nowhere in the Occupational Health and Safety Code does it require that safety meetings be dry, dull, and boring. Safety isn't boring in and of itself. It's the presenters who are boring. A good presenter can bring interest to the safety discussion and even make something as mundane as air particulates interesting and engaging. Conversely, a bad presenter can make any subject boring.

With so much focus on overcoming high disengagement rates, safety people and supervisors need to up their game at meetings. If you want your people to make good decisions in the field, then you've got to ensure that the information you communicate is heard and understood.

If you want to become a safety leader, your focus has to be on engaging your people, not on achieving the bare minimum in safety. It's one thing to say you gave your people the information at a meeting. It's quite another to ensure that your people understood and internalized that information in a way that inspired them to perform better with safety.

LEARN TO LEVERAGE THE GATHERING

People who aren't engaged aren't listening. If they're not

listening, you're talking to yourself. You wind up preparing a presentation to deliver to a figuratively empty room. If the information presented is not understood, you won't even achieve compliance. If you believe that the information is important and people need to know it, you must first ensure they are paying attention. Once you have their attention, you have more assurance you will be heard.

Boring meetings are boring by choice. To attendees, it speaks volumes when the presenter doesn't care enough to make the meeting at least halfway interesting. To impart information, a safety-meeting chair must first engage the audience.

Safety meetings should be used as an opportunity to motivate and inspire, an opportunity for employees to ask questions and express ideas. One-way lectures are rarely engaging, unless the presenter has superior presentation skills. And even if that's the case, safety meetings, by their nature, should be participatory. Presenters need to ensure that their audience understands the information. You don't want passive attendees; you want engaged people.

Think of your safety meetings as a pregame meeting. Teams heading onto the sports field don't spend their time reviewing how to fill out forms. Coaches use the opportunity to uplift every member of the team to ensure they're all mentally prepared and focused on the task at hand. Coaches want to get their players to buy into the game plan. Safety meetings should also have a game plan. There should be one main thing that you want your crew to focus on, and you need to get them motivated to do it.

TREAT IT LIKE AN INVESTORS' MEETING

If you've ever sat through a timeshare pitch, then you know that a good presentation speaks directly to you. There are no slides featuring pictures of a brochure. There's one or two people at the front of the room who are there to make a compelling argument about why you should invest. They want your money, and they will make it very difficult for you to say no. These presentations are engaging. They force you to picture yourself enjoying a timeshare for the rest of your life. They make their pitch feel personal. It's as if they're speaking directly to you, and only you. It's compelling and very tempting.

On TV shows such as *Shark Tank* in the United States and *Dragon's Den* in Canada, entrepreneurs pitch to investors. They're looking for a buy-in. If the pitch misses, there's no investment. The same thing happens with safety meetings. If your pitch misses, your people won't invest themselves in safety. Your presentation won't yield the buy-in you're looking for.

So take an honest look at your pitch. Study the last safety meeting presentation you made. Was anyone convinced to invest their time, energy, and effort in your goals for improving safety? Did you prepare compelling arguments for investing in safety? Ask yourself whether reading a bunch of stats and data created a groundswell of safety investors. Did your pitch convince your best people to buy into the latest initiative?

People will buy into anything that creates a win for them. When people can clearly see the personal benefit, they're more likely to buy in. If you use your safety meetings the right way, you remove the mental barriers to investing in safety.

Timeshare pitches remove your reasons and excuses to say no, one by one, until you are left with nothing to stand in the way of a timeshare—not even financing. In the same way, your safety pitch has to remove the barriers to a buy-in by your crews. Make safety an appealing investment instead of a stuffy set of rules to follow.

MEETINGS SHOULD BE CONVERSATIONS

A long, boring lecture makes it tough for people to stay focused and attentive. After a while, their brains will simply check out, and the presenter will be left essentially talking to no one. If you want people to participate, make the meetings participatory.

Supply pens and paper and ask attendees to take notes and write down questions. Then open the floor for a Q&A session. Present certain questions of your own and solicit answers and opinions. Present debatable topics. You can even take a vote on issues. Ask for new ideas and suggestions for how to improve parts of the safety program. You can relate a recent incident on the job and ask attendees to make suggestions for how it could have been prevented or corrected.

This is all very different from expecting people to sit there while you drone on with your PowerPoint. If the power went out, could you carry an hour-long meeting without it? Safety meetings need to be more than how many boring slides you've prepared. Safety meetings need to inspire more than they need to inform.

However, if you're afraid of being shown up by employees, you'll avoid asking their opinions. You won't ask for their involvement. If you're afraid that employees might offer something you didn't think of, you're probably guided by ego or insecurity, not real safety concerns. Safety meetings then become all about you and not them. If the safety meeting isn't about helping employees, they won't buy in.

In order for safety meetings to work, they need to be inclusive. That's the bottom line.

MEETINGS ARE FOR REACHING PEOPLE

Independent Well Services (IWS) in Estevan, Saskatchewan, supplies service rigs to oilfields in the Bakken Formation Region. IWS operates a fleet of nine class II freestanding mobile service rigs. Although IWS employs only around sixty people, it packed the Estevan Memorial Hall with nearly three hundred attendees for its 2016 Annual Safety Day. They invited customers, clients, and their direct competitors to attend.

Brian Crossman, field supervisor and one of three founding partners of IWS, handles sales and marketing for the company. He explained that the purpose of the meeting was less about "how you do things, and more about how you think about things." The meeting focused on soft skills rather than technical training.

Estevan is not a terribly large region. It's a pretty tight-knit community, especially in the oil and gas business. Crossman opened the meeting to as broad a base of people as possible because he wanted to reach the widest audience about safety. "I know all the people who run competing companies. I would hope that if we needed help one day that they would be there for us. I thought it was just the right thing to do. I would hope that maybe one day down the road they might do something like this and invite us. Some of the guys who currently work for our competitors might one day come over to work for us. They will already have gotten the right information about how we work with safety."

Crossman and his company illustrate an important point about safety meetings. The primary goal is to reach people and have a positive impact. Safety meetings aren't about trade secrets but about best practices and buy-in. IWS wants to make the best information available to the most people. After all, safety is everyone's concern.

ALIGN SAFETY WITH YOUR WORKERS' PERSONAL VALUES

You can't create a safety culture through just compliance measures. When employees discover that the company shares their values concerning safety, they are more willing to buy in. Safety has to be personal. People need to feel that what they believe and what the company believes are the same thing. Safety culture results from the collective attitudes in a workplace. People tend to respond more favorably to safety when the safety program is aligned with workers' personal values.

Safety meetings can be utilized effectively to establish common ground around safety values and to rally everyone to a common cause.

You can start to do this by taking existing corporate values and aligning them with employees' personal values. This requires plenty of conversations, both one-on-one and in a group setting. Conversations, not lectures. Take the time to understand each employee's values and goals. You can use meetings to solicit ideas, thoughts, and opinions.

When you see how crew members' values and your values align, you'll be in a better position to show them how safety can get them to where they want to be. You can use meetings to build a consensus of shared goals for what people want out of safety.

HOW TO SPOT NONALIGNMENT

After a day of freezing rain followed by two days of snow in Regina, Saskatchewan, I sat at a stoplight. I'd just delivered a safety leadership presentation. In front of me, I saw a car fly through the intersection with its windows, side mirrors, and tail lights completely covered in snow. The driver turned into a construction area, backed his vehicle into a parking spot, and stepped out of the car in full safety gear: hard hat, safety gloves, steel-toed boots, and coveralls. He was dressed for safety in the workplace, but it was clear he hadn't bought into safety at the personal level. He was the kind who tolerates safety rules while at work, but doesn't believe strongly in safety on his personal time.

The simplest way to spot this kind of nonalignment of workplace safety and personal safety is to look for contradictions. The guy who speeds to work, or has burned out brake lights, or wears flip-flops when he mows the lawn isn't going to come into work and suddenly be Mr. Safety. He might tolerate the rules, but he hasn't bought into safety as a personal value. If all you're looking for is rules compliance, you might be happy with employees tolerating rules. However, if you want to build a solid safety culture, you're going to need employee buy-in.

Supervisors and safety people can create nonalignment by being overly focused on enforcing rules. That's the cop or security guard mentality, and, sadly, that's how many safety people view themselves. It's a mistake to use safety meetings to play cop. Employees resent being policed. It's no way to win them over at a meeting. If all you're focused on is enforcing rules, you'll only foster resentment among your crew.

You build a culture of safety by aligning personal and

company values with a far-reaching commitment to safety at all levels. Not all employees possess the same level of knowledge, nor do they learn at the same speed. Each employee has their own set of values as well as cultural and educational backgrounds. At meetings, it's therefore important to involve each individual. Encourage each person in attendance to get involved and to stay involved. Your job is to personalize safety for each employee.

Meetings are meant to inspire and motivate people, not to bog them down with rules.

FIVE STEPS TO THE PERFECT SAFETY MEETING

There are many factors involved in safety meetings, but what's most important is how safety meetings are used. Traditionally, safety meetings are treated as information dumps. This is wrong from the get-go. Safety meetings should primarily be used to motivate, and there are many ways to do so.

What follows are five primary ways to overhaul your safety meeting format. Each of the steps below is designed to snap you out of habitually dull meetings.

Einstein said, "We cannot solve our problems with the same thinking that we used when we created them." Sometimes a change in venue can give our safety meetings a creative boost. So instead of holding meetings in the dirty back shop, try taking your crew out for breakfast. This can breathe new life and ideas into the safety discussion.

Doing something unexpected can help you elevate the perception of the importance of safety meetings. If you can't take your people out for breakfast, at least bring a meal in for them every once and awhile. This can snap your crew out of their normal mental checkout and break down their resistance to another boring safety meeting.

When people can relax in this way, it encourages conversation. This can make asking a simple question end with a flurry of ideas. Ask simple questions, such as: How can we improve our safety program? This can spur an hour-long discussion. The challenge is keeping the focus of the meeting on solutions instead of complaints about what's wrong. When crew members provide solutions, write them down. Let your people know that their input really matters.

2. SHORT, SWEET, AND RELEVANT

TED Talks is a worldwide speaking series that invites thinkers and doers with leading-edge ideas to share their life-changing messages. TED stands for Technical, Entertainment, and Design, and TED presenters have to get right to the point of their message. This is because they only have eighteen minutes to present a concise and compelling pitch.

If world-class thinkers can communicate their life-changing messages in under eighteen minutes, why do we need to give safety meeting presenters thirty to sixty minutes to explain the right way to use a ladder?

When the length of the scheduled meeting becomes more important than the content itself, you're off track in making a compelling case for safety. If you give a presenter thirty minutes to cover a topic, they'll stretch their subject out. They'll pad and stuff their presentation to fill the entire thirty minutes when the subject could probably be adequately covered in fifteen.

Instead, don't force presenters to fill time. Schedule them for shorter time periods to present interesting and topical subjects highly relevant to the work. If it's December, then your presenters might cover winter slips and falls. If it's July in Las Vegas, heat-related issues would be more practical and valuable. Focus on the issues that are most relevant to your people. Ask them what they want to know more about and plan the next meeting around one of those subjects. Keep presentations short and relevant.

3. MAKE SAFETY ABOUT POSITIVES

Safety is supposed to make lives better, but traditionally it hasn't been presented that way. The main focus of safety has been what you can lose if you don't do it right. You could lose a limb, a life, your income. Traditionally, this negative approach has been used to scare employees into conformity and safety rule compliance. However, safety is not supposed to make you afraid. It's supposed to make you confident in your ability to make good choices and to get right outcomes. Safety should give your people the opportunity to excel, exceed, and drive good results. When you can inspire people to value and trust safety, then you create a win-win scenario where safety is the answer.

4. THINK OF SAFETY MEETINGS AS TEAM PROJECTS

Organizing and running exceptional and engaging safety meetings usually exceeds the skill set and time constraints of most safety people. They may know the safety regulations, but that doesn't mean they're exceptional meeting planners. Besides, safety people also need to be out in the field. They don't have an unlimited amount of time to dedicate to organizing and pulling off memorable safety meetings. Someone needs to help out. It should be someone who can organize inspiring events, create engaging subject matter, and figure out meeting logistics. This usually isn't the domain of safety people.

As an organization, you want to make sure that there's somebody working with the safety department who has these skills. Safety is a team project that requires a project manager. The project manager doesn't have to be a safety person, but rather someone with organizational skills. Get the meeting part right and you'll get the safety buy-in you need.

5. HAVE A CALL TO ACTION

Safety meetings can have one or more objectives. Primarily, they need to solve a problem and make your organization better. This requires you to consider the question: How can I know if I had a successful meeting?

The simplest way is to ask your people to do something. Set expectations and then end your meetings with a call to action. Make it clear to meeting attendees what you expect of them coming out of the meeting. The purpose of the meeting is to align people with an action and result.

Whether it's a toolbox meeting or an annual safety event, every meeting should have a call to action that either advances an idea or inspires people to be better.

BUILD CONSENSUS

Safety meetings are opportunities for open dialogue and a free exchange of ideas. They are forums to build consensus. When people agree on basic safety principles, it's more likely that they'll buy into new initiatives.

Meetings can be used to help reinforce the crew as a team. When crews become part of the discussion and conversation about improving safety, they are that much more invested in building a culture of safety on the job.

EMPLOYEE SURVEYS

· · · · · ·

It's not uncommon to survey employees about what they would like to cover at the next meeting. Surveys can also be used to find out what attendees thought about a previous meeting. How effective was it? How effective was the call to action? In this way, you can find out how employees are using what they learned.

CHAPTER

MEETINGS: CREATING EMPLOYEE BUY-IN

Simon Sinek, optimist and author, once said, "People don't buy what you do. They buy 'why' you do it." Sinek was referring to his Golden Circle model of codifying the world's great leaders. In Sinek's Golden Circle, there are three concentric circles. The middle circle is the "why" of what we do. Outside that circle is the "how" circle, representing how we accomplish what we do. Outside of those two circles is the "what" circle. The reason "why" is in the middle is because it's central.

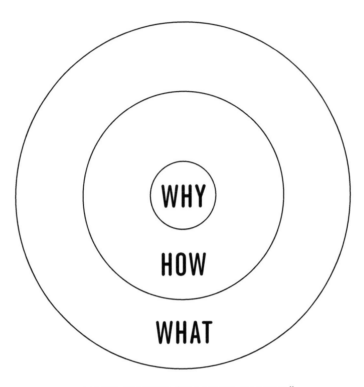

WHY IS "WHY" SO IMPORTANT?

When asked what he does, a safety person might say, "I'm a safety advisor. I ensure that workers don't get hurt on the job by communicating rules, doing inspections, and enforcing the Occupational Health and Safety Code." This explanation contains "what" he does, followed by "how" he does it, but it's missing the most important part, which is "why." In fact, the "why" of safety is the most compelling part of the story.

"Why" is the question children ask most. Why? Because it's the most interesting thing to know.

So instead, when asked about his work, the safety advisor could say, "I work with a professional team of talented people who build some of the most outstanding buildings in our city. I really believe that their talents are indispensable and that their contributions need to be protected. So I help my employer protect these valuable assets."

That's the "why" right there.

He could go on to say, "Here's how I do it. I engage my fellow workers and communicate and inspire them to make good choices for their own health and safety. I help them be mindful of their tasks and the people standing next to them. It's vital that they really understand and appreciate how valuable they are to our organization."

That's the "how" of his work.

He might conclude, "I'm a safety advisor on a commercial construction site, and it's very rewarding work helping my fellow workers build good lives for their families."

That's the "what."

Same job, but explained first by "why," then "how," and, finally, "what." The "why" of his work is most powerful, and it proves Sinek's assertion that people don't buy what you do; they buy why you do it.

So let's talk about why your people should buy into your company's safety program and how you can use safety meetings to deliver that. And remember your "why" will be a key component in helping them find their own "why."

CHANGING THE CONVERSATION

We've discussed how safety has both long- and short-term consequences. Safety has traditionally been focused on pointing out what workers could lose if they choose outside of safety: a limb, an eye, their life. These gruesome images and threats are found more inside safety meetings than anywhere else in the safety program. The threat of loss is thought to be an effective motivator, one that is supposed to spur compliance with rules. Unfortunately, negative reinforcement as a motivator has no lasting value and requires constant repetition, each time ramping up the degree of shock in order to make an impression. Still, the best you can hope for from this negative reinforcement is simple rules compliance. However, rules compliance is very different from a true buy-in to safety. Compliance is based on the enforcement of rules. Buy-in is based on embracing safety as one's own personal value.

Therefore, the argument has to change. This requires changing the conversation. At safety meetings, instead of focusing on what your people might lose, focus on what they'll gain. Put a positive spin on safety to show why it's good for them. In fact, there's a simple strategy for achieving just this. It's called the DSL Strategy.

THE DSL STRATEGY

The first part of this strategy, the D, is Desire. What do your employees want in the short term? Whether it's a big-screen TV, a new car, or paying for their child's education, desire is all of the things that depend on having accessible cash flow in the immediate present.

The ability to pay for one's immediate desires without the burden of going into debt requires ongoing income. Life is expensive and options are limited. Taking unnecessary safety risks jeopardizes continuous cash flow and, therefore, fulfillment of desire. Doing risky things off shift can be even more consequential than taking risks on the job. When you're off shift, there's no insurance or Workers' Compensation. When you're not working, cash flow stops. So choosing safety assures cash flow and the good life.

The second part of the DSL strategy, the S, is Security. Employees don't ever want to be a burden to their spouse or kids. They don't want to experience a financial crisis that forces them to downsize their lives. By choosing safety, employees can plan for the future and ensure their own security. Choosing safety provides long-term security for families and reduces stress.

Part of that security focus involves a secure retirement. When you ask people what they plan to do when they retire, everyone has a story. It might include buying an RV, moving to a cabin in the woods, or a beach house. It might even involve world travel. Everyone has some sort of vision of retirement. Moreover, because they have a vision, most people put money away each month to make it happen.

But ask them if they have a plan for living long enough to reach retirement age and most won't understand the question. Most people live each day without thinking about how it will affect their retirement plans. Simply put, not every person makes it to retirement age. My own father died three months short of sixty-three, after a lifetime of smoking and alcohol abuse.

Embracing safety and healthy living as a personal value is an investment in the present that could reasonably extend your life.

If you're a safety manager or supervisor, meetings are forums to show your people how the safety program is a plan to make it to retirement. The safety meeting is a great place to have an informal conversation about how decisions yield results one way or the other. Encourage your people to tell their "today" story at safety meetings. Help them to be proud of the investment they are making in their long-term security. Just like putting money away each month for retirement, your people can make smart decisions each day to reach retirement age safely.

The third part of the DSL Strategy, the L, is Legacy. In twenty years, if you are asked to give a speech on safety, would you want to tell about how safety gave you a good life and good health, or how ignoring safety cost you a healthy retirement?

Employees need to know that their legacy starts today and that each day they build on it. Legacy starts with choices, and choices yield results for good or bad. It's their choice. Safety meetings are opportunities to lay out the choices and show the path to a positive legacy.

MAKING CHOICES

As a young worker, Jeremy is a framer for a homebuilder in Red Deer, Alberta. Like many young workers, his excitement at starting a new great-paying job overshadowed the potential risks on the job site. He paid more attention to the earning potential than he did to the safety training he received.

We had lunch together one day. The conversation turned to a discussion about working at height without fall protection, his right to refuse unsafe work, and a host of other safety issues and concerns. He was unaware of his own responsibilities in safety. To him, safety rules and regulations got in the way. Like many young workers, he talked with an invincibility mindset.

"What happens if you get hurt?" I asked him.

"Won't happen—and if it did, she'll take care of me," he said, referring to his girlfriend.

"Have you asked your girlfriend if she's okay with looking after you for the rest of your life?"

There was an uncomfortable silence. His girlfriend was in college studying for her own career. She had hopes and dreams for their lives together. I asked him what their future would be like if he were seriously injured on the job and then became a burden to her. How fair would that be to the girl he loved?

Every day we make choices on and off the job, and these choices should be discussed at safety meetings. Choices about alcohol, drugs, cigarettes, or even being overweight can adversely affect an employee's job performance and long-term health. There are plenty of statistics to reference, but more important is the personal side of things such as Jeremy's choices.

At meetings, we don't want to single out individuals as bad examples; we want to bring attention to positive examples. We want people to learn from the positive choices their co-workers make: the guy who quits smoking, the framer who avoids risk and can explain how he does it.

OTHER WAYS TO GET EMPLOYEE BUY-IN

A one-hundred-meter sprinter works every day to knock one-tenth of a second off her best time. Why? Because it's enough to make the difference between winning and losing. Competitive athletes work on skills and techniques to maximize their potential. Much of that fine-tuning takes place on the mental level. It's no different in frontline work.

Safety people need to turn the focus of safety meetings to techniques that will fine-tune performance. Supervisors should bring attention to workplace safety skills development and improvement, and some of that improvement occurs at the mental level.

Teamwork involves a mental attitude that is brought to the field each day. Just as a basketball, soccer, or hockey team drills their plays, safety meetings can develop group dynamics, awareness, and cohesion that workers bring to the field. Crews are team players who need to look out for each other. They need to coordinate and optimize their moves for maximum effect. They need to play and win as a team. This is where you as a safety person or supervisor can help your team prepare their mental safety game.

As a supervisor or frontline safety person, you can help employees remove the mental barriers to safety buy-in. In safety meetings, address the reasons and excuses that prevent employees from choosing safety. Hash things out in open discussion and encourage dialogue. Facilitate safety meetings that allow attendees to point out when their co-workers are making excuses for not embracing safety. Have honest discussions about how they can help improve safety.

When supervisors lead meetings with honest, frank, and open discussion, attendees respond to the process. They want to feel empowered and are even surprised at their own willingness to speak up. You can help unlock their personal leadership potential. This is a strong concept to consider employing as part of your safety meetings.

MAKE THEM FEEL IMPORTANT

Back when baby boomers were the dominant generation in the workforce, motivating them was simple. Managers simply threatened to fire them, and out of fear, employees would get their act together and buckle down to do good work. Today, though, threatening to fire an employee doesn't work. Threaten to fire an employee, and they're likely to quit on the spot.

From an employee's perspective, if you hang the job over their heads, you don't value them. Nobody wants to feel like their contribution isn't valued. No one wants to feel less important than someone else. People ultimately want to feel that their contribution matters and that they are valued.

Codesafe Solutions Services in Melbourne, Australia, understands the power of making people feel valued. In fact, their corporate mantra is "Valued people value safety." The company takes complex safety manuals and turns them into short videos. These convenient learning tools can be accessed by employees on mobile devices right at the point of task.

According to David Broadhurst, the company's cofounder and CEO, "Paperwork and documents make workers' eyes glaze over. Using documents to communicate critical information to guys with trades backgrounds misses the point. Most trades people have learned by watching, not reading. So Codesafe provides visual refresher training to increase knowledge and performance."

Broadhurst says he was brought up in the construction industry. "I've been around pipelines and heavy construction all my life. As you move through the ranks of labor to supervisory roles, you start to see a pattern in the types of workers who seem to take more risks than others. When I was given a supervisory role on large projects, I began to see how I could influence behavior. What it came down to was making guys feel valued."

He explains that when we value the people we supervise, there's less of a chance they'll do something risky to themselves or others. "If you ask workers for their help in improving processes and procedures, they'll gladly offer ideas because they want to feel valued and heard."

There are two basic styles of leadership. "You dominate or you influence," says Broadhurst. "In trades-based industries, we think we're very good at the first. We think we get results while we're there, but as soon as our backs are turned, there's rebellion. So the question becomes: As a supervisor or manager, how can you influence and not dominate?"

Codesafe Solutions brings their "Valued people value safety" philosophy to the companies they serve. "It changes behaviors and improves outcomes basically overnight. Organizations just can't believe how fast we can change the culture and the business. We make it about valuing and honoring people. As a manager, you are the facilitator of a process. It's not a top-down approach. It's an equal responsibility approach. Management has to relinquish some of the control, but they can still craft the conversation where they need it to go," he says.

"We give our clients a framework for how to engage workers. They give the workers the autonomy they need to take part in their own destiny. It's not that complicated. It's the basics of human behavior. It's how we make people feel valued."

HOW A SUPERVISOR IS A MODERN-DAY COACH

If you're in a position where you oversee others, but don't have the rudimentary skills to effectively manage or coach them, then you need to rethink your own value to the position. Smart safety people and supervisors are making the shift from enforcement of rules to coaching behaviors and influencing decisions. The objective is to inspire best performances that become habits. In safety meetings, it takes very little effort to read off a list of inspection and incident reports. However, it takes much greater effort to use these moments and reports to inspire better performance.

It's what any good coach does. Coaches give their players small achievable goals to attain. They focus on one small improvement in performance at a time. They don't follow players around criticizing.

The point of becoming a coach is to help others improve by giving them a sense of self-worth and personal value. Your goal is to communicate small bits of knowledge and experience to make your people better. The point is to build your people up, not tear them down. No one ever improves by being torn down.

You can be either a coach or a referee. Referees enforce the rules, but they don't help players improve. Coaches build their players' skills, improve their performance, inspire and motivate, and help their players play better within the rules. Are you a rules enforcer or are you willing to do what is necessary to help your team improve their performance?

When a whole team continually underperforms, who's to blame? Here's a hint: it's not the players. When the whole team underperforms, the coach is the one who gets fired.

You never see a whole team get fired. Instead, management looks for a new coach—one who can bring a new approach that the players can buy into. A good coach who values his or her players usually gets better performance from the same team.

So ask yourself about your approach. You can tell a lot about the effort and willingness of safety meeting organizers by the way in which they put the safety meeting together. Half-hearted efforts in approaching safety meetings are more likely to result in half-hearted efforts in safety performance. If you're not willing to put in the effort to conduct engaging safety meetings, you are telling your people that you can't be bothered, and they won't be bothered either.

Just as a coach has to know each one of their players and what makes each one tick, you should also know your people. Everyone learns differently. Some are more visual, others are more auditory, and many more learn by doing. You need to know how each person learns so that you can make safety personal to them. As a supervisor, you're probably dealing with a crew of six to eight, so take the time to get to know each one. Figure out how they learn and the individual parts of their performance that they may need to work on. Make the safety meetings specific and personal for each individual.

People are only able to achieve excellent safety performance when they receive excellent coaching. People are uplifted by uplifting meetings. Are your safety meetings uplifting? When employees feel that safety meetings are events for just reviewing paperwork, how are they supposed to look forward to them?

When safety meetings are ambiguous and scattered, employee focus and attention is scattered. If your people aren't focused, they won't perform well. Let your people know in no uncertain terms that you value and care about their contributions to the team. Ask for their input at safety meetings. Engage their intelligence and experience, and watch how they come around.

BE PART OF THE GREATER GOOD

There's a growing trend that shows that traditional companies are turning their backs on bare-minimum-compliance safety programs and meetings. They are changing the paradigm within their own companies. They are upping their own corporate standards to achieve safety at a higher level. When the expectations of managers and supervisors have been raised, the expectations for the kinds of safety meetings they conduct has to follow.

More and more companies are making safety a key cornerstone of how they do business. You only need to attend a company's safety meeting to be able to determine the kind of safety culture that company has. If you want to keep pace with the market, you've got to be onboard with how you conduct your safety meetings. Bare-minimum safety is only for companies prepared to go after a shrinking market of table scraps. However, companies focused on integrating safety into everything they do are starting to compete at a higher level and for a bigger piece of the pie.

If you want to play with the big boys, then your safety attitude needs to be set to maximum. That includes how you use your safety meetings to rally your people around your safety culture.

CHAPTER

MARKETING: SELLING SAFETY

If you think that marketing is all about sales, advertising, and late-night infomercials, then this chapter should change your mind. Marketing is the third component of the M4 Method of building solid safety programs.

For supervisors and safety people who want to get important messages to employees, marketing is a way to communicate the "why," the meaning and importance of safety, to frontline staff.

MARKETING IS THE "WHY" OF ACTION

A sign in the workplace that reads, "Stop. Do Not Proceed," is a warning and nothing more. It's a type of communication, but it doesn't communicate the "why." Memos, PowerPoint slides, emails, toolbox talks, and safety training are all forms of communication. They communicate information, but do they also communicate the inherent value of safety? The "why" of safety?

Safety people need to take communication to the next level. We need to create communication that motivates employees. This is where marketing comes into play. It's why marketing is the third critical component of the M4 Method.

Marketing moves people forward. It's a spur to action that takes people from where they are to somewhere better. Marketing goes beyond simply informing. It's the next stage of communication. Marketing makes a compelling case for "why" employees should choose to take a new action for safety.

Simply put, marketing is what creates value and motivates people to action.

MARKETING SAFETY TO HEARTS AND MINDS

In order to market something, you have to appeal to the hearts and minds of your buyers. Unfortunately, safety has traditionally focused on appealing only to minds and not hearts. This is why safety buy-in meets resistance: there's been no establishment of value for the buyer.

To understand how to better position safety as something of value, first, you have to understand how people buy. People don't buy things logically. They buy things emotionally. Once they have made the decision to buy in, people then use logic to justify their purchase. They'll use statistics, facts, and figures to validate the decision they made.

Let's say you're interested in buying a new truck. You look at a brochure, you read the stats, check the fuel economy, the warranty, and reviews, but that isn't enough to make you buy the truck. Eventually, you'll go to the dealership, sit in the vehicle, smell that new car smell, and put your hands on the wheel. Then you'll take it out for a test drive, put it through the paces. Throughout the entire hands-on experience, your brain is in overdrive, processing information in combination with sight, smell, sound, and touch. How does it make you feel? That's the emotion kicking in, and that's what sells you on the truck.

Once you've made your emotion-based purchase, you'll justify it with logic. This happens when you buy a new house, choose a vacation, purchase new clothing, or nab that new cell phone. It's a process of emotion informing a buying decision.

It's this emotional component that workplace safety hasn't quite figured out yet. Only a few companies are beginning to understand that people want to be part of something bigger that makes their lives better.

Workplace safety can in fact bring people together in an emotion-based value system. When done right, marketing can help turn a company's mundane safety program into a movement built around a set of values rather than rules.

THE TWO SALES OF SAFETY MARKETING

Back in 1997, in a moment of weakness, I did something that I'm not too proud of—I became a salesman. In fact, I spent several years selling photocopy machines, as well as radio advertising, before I opened my own business. What I learned as a salesman, a manager, a business owner, and an entrepreneur is that selling is largely about selling yourself, your ideas, and your point of view.

While I was selling photocopiers and fax machines, I discovered that I actually preferred training salespeople to selling the machines. I spoke to corporate senior management about that possibility and made my pitch to take on the position. My immediate manager didn't much like the idea of me leapfrogging him in rank, so he gave me an ultimatum: I had six weeks to sell an additional $60,000 in equipment, or I'd be fired. I turned in my notice that same day and spent every free minute of those six weeks writing what would be my first book on sales training.

In that book, I began working out the Two-Sales Rule. In every sales situation, there are always two sales that need to be made. The first sale is you: the salesperson. You have to successfully sell yourself as a trustworthy person who can provide something of value. The second sale is whatever's in your briefcase: the product, service, or point of view. If you don't make the first sale successfully, then you're less likely to make the second one.

SAFETY MARKETING: A BALANCING ACT

Marketing safety is a delicate balance of personal inter-action combined with creating value. When you boil it down, marketing safety is ultimately a series of sales calls, so the Two-Sales Rule applies. First, you've got to sell yourself. Once you've established yourself as cred-ible and trustworthy, a person others like to work with, then you can move to the second sale, which is selling the safety message.

To most supervisors and safety people, the idea of selling safety is vulgar. Those who have strong beliefs about safety think that safety should sell itself—that people should buy in because it's the right thing to do. However, we know that people don't always do the right thing. They smoke cigarettes, text while driving, drink to excess, and overeat. People sometimes need a little convincing to make changes in their own best interests.

Sometimes that convincing involves personal interaction with someone they trust. Safety marketing is easier when you become an influencer first. Again, it's the Two-Sales Rule. Once people have bought into *you* as someone whose influence they trust, then selling the safety message becomes that much easier. When people see you as trustworthy, they're more willing to listen to you. You can then use your influence to market the safety program as something of value to their lives.

MARKETING SAFETY IS A GOOD THING

Historically, safety has been treated as a compliance measure with rules, processes, and procedures for ensuring that no one gets hurt. However, safety actually works better and is more successful when the team actually buys into it as a value they care about. People will buy into safety when they see what's in it for them.

People don't like the hard sell. They don't like to feel manipulated. They don't want to buy something that doesn't appear to have value to them. They don't want things shoved at them. People are resistant to that approach, so hammering on safety rules or acting like a safety cop won't sell your people on safety. That isn't good marketing.

However, people love to buy something they value. Safety people, therefore, can begin to get buy-in when they successfully communicate the "why" of safety. When they outline safety's benefits in such a way that each person on the team understands "what's in it for me," supervisors can begin to successfully market safety.

ESTABLISHING THE WIN

As noted at the beginning of this chapter, safety has overwhelmingly used warnings as a communication strategy. Most of the emotional appeal of warnings is focused on creating a negative emotion, including sad stories of injury on the job. Gruesome photos of dismemberment create emotion, yes, but not the kind that moves people *toward* something.

Warnings don't work as marketing. The words "don't," "never," and "stop," aren't good marketing messages. "Don't do what I/he did," doesn't provide positive value that leads to positive choices and results.

As a supervisor or safety person, you need to know the answers to the following questions:

- What are the positive aspects of safety?
- What are the benefits of doing safety right?
- What's the win-win of buying into the safety program?

When you can answer these questions in a way that establishes value for employees, you'll be on your way to selling safety successfully. You'll be prepared to find the win in safety for each of your people—the answer to "what's in it for me."

SAFETY IS A COMMODITY

You may still be thinking that safety should sell itself; however, in the marketplace, few products sell themselves. They all need a push. Companies spend money and time marketing their products and services in hopes of developing new customers. They market to job seekers and their own employees about the company's successes and achievements. They even market to their own industry to demonstrate their leadership.

A company that can market its products and services to strangers in a compelling way can also market its safety program to its most valuable people—its employees.

People will gladly buy into something when they can clearly see the benefits of owning it. They buy in when they're convinced that doing so will make their lives better. Safety needs to be marketed in such a way that employees choose safety. In this way, safety becomes a commodity they want as their own. They feel motivated to take ownership of the safety program.

GIVE PEOPLE WHAT THEY WANT

To overcome the mental barriers that employees may have toward buying into safety, supervisors need to shift the crew culture one person at a time. This is very different work than trying to convince somebody to comply with procedures. Selling is helping people get what they want. So what do people want?

Most employees want to feel valued for the work they do. They want to feel like their work means something and that their contributions are important. They want to feel protected by their employers and respected by their co-workers and supervisors. In a nutshell, people want to feel proud of what they do and how they do it. They want to be admired for their expertise and teamwork. Above all, they want to enjoy and experience the goodness and richness of life without having to fear for their safety or security.

The good news is that supervisors and safety people can deliver that. Supervisors and safety people first have to reframe how they market safety to employees. They must do it in a way that shows employees that safety and caring for people gives them exactly what they want.

INFLUENCE VERSUS MANIPULATION

As mentioned earlier, people don't respond well to manipulation. Incentives that involve elements of fear or guilt to achieve basic minimum compliance aren't effective.

The website EmergingLeader.com has this to say about manipulation compared to influence: "Manipulation deliberately uses and abuses other people to act out your intentions. Influence, on the other hand, requires buy-in on the part of the person being influenced and the willingness on their part to support your goals."

Safety leaders should also steer clear of admonishment. People can't be scolded into safety. Admonishment is a clumsy tool used by underskilled supervisors and managers. On social media, it's not unusual to see supervisors and safety people publicly admonishing their own senior executives for their perceived lack of commitment to safety.

That's no way to encourage and influence buy-in at any level. Singling people out and scolding them in public is no way to build a solid team of performers.

BRAND IDENTITY

Marketing builds on positives, not negatives. Successful marketing creates a bond between the seller and the buyer. Just think of how attached people get to their cell phones, their brand of computer or car. Personally, I'm a Blackberry man. I love my Toyota RAV4 and my Lenovo ThinkPad. I also happen to be a "safety" man. I've bought into safety and own it as one of my personal values. Safety is my preferred brand. People buy into safety when it offers them an identity they want to be associated with, like a cool gadget or power tool.

Safety isn't just something we do. It's not an add-on. It's the foundation on which an organization is built. Safety needs to become one of the core values of a company alongside accountability, responsibility, and integrity. That's why safety marketing can't just focus on safety. It shouldn't just change safety culture; it should change corporate culture. It needs to be integrated into all aspects of the company. In this integrated approach, safety becomes one of the foundational values of an organization.

Safety marketing creates a positive association with safety. It brings safety front and center. Marketing safety, when done right, brings the message home. Safety needs to find its way into the hearts, hands, and minds of crews, customers, and contractors. Everyone associated with the company in any way needs to measure him or herself against that message to see how well he or she fits with the organization.

For instance, PENTA Building Group in Las Vegas holds a big safety event every year. They invite contractors, subcontractors, clients, and competitors to attend and celebrate safety. Each person who attends gets a jacket with the crest of the year's safety theme. Those emblazoned jackets become walking promotions for safety. In this way, PENTA's people take their safety program outside the company's walls.

Another example is the North American energy producer Encana Corporation. Their marketing program is called Courtesy Matters. As most of their work is done in rural communities, Encana wanted to tie together company culture, safety, and community relations into a campaign that demonstrates its good neighbor policy. When driving down a dirt road, employees will slow down to avoid raising too much dust for the people living close by. They'll close any open gates they see. They follow company policy by doing what any good neighbor would do, and their respect and courtesy spills over into safety. Employees willing to be courteous with neighbors are more apt to display that same courtesy for co-workers.

To market safety successfully, you must connect employee pride to the message. If you can make people feel proud of where they work and of the work they do, then they're going to engage. Are your employees proud of your company? Do they list your company as their employer on their Facebook page? Employees are more enthusiastic about flying the company flag when they feel genuinely proud of where they work. If your company is Facebook worthy, then harness that pride. It will increase employee willingness to buy into safety.

MARKETING SAFETY 365 DAYS A YEAR

Each day, week, month, and year, a good marketing campaign continually puts out the safety message. The message is refined and improved to better drive home the core message of safety. We should always be finding new ways to connect corporate safety goals with personal values. Safety marketing aligns every employee as an important part of an organization's success and reaches out to families and communities.

In order to sell a product, philosophy, or idea, you must have the attention of the listener. Without engagement, the message is not reaching its target. That is one of the key components of selling a message of safety. More importantly, in order to continue to be able to sell your message, you can never stop broadcasting. When you stop broadcasting your safety message, you allow competing messages to have a voice.

Safety needs to take a lesson from basic marketing: What gets focused on gets talked about, and what gets talked about gets attention.

The launch of the latest iPhone or Tesla car model gets attention. Media talk about it. People search online for more information. They even stand in line at retail outlets to be the first to buy. Imagine that kind of reaction to your latest safety initiative. Your job is to build excitement about safety, get buy-in, and keep safety front and center at all times.

In the next chapter, you'll learn the nuts and bolts of how to build a successful safety marketing campaign. You'll also learn the strategies and stages of how to put marketing to work for safety's sake.

MARKETING: HOW TO BUILD A SUCCESSFUL SAFETY MARKETING CAMPAIGN

Warnings warn. Communications inform. But marketing moves people. To the supervisor or safety person, marketing safety may seem foreign, even unnecessary. Done right, however, safety marketing streamlines rules compliance, reporting, paperwork, and buy-in to safety. When done effectively, safety marketing helps improve safety performance.

Marketing is about building value. Safety marketing is about helping supervisors and safety people connect the value of safety with the needs of the employee.

Every employee has needs: food, shelter, family, quality relationships, gainful employment, pride in work. The task of safety marketing is to establish how safety can satisfy any or all of these needs. When safety marketing is done successfully, employees make the right connections between safety and fulfillment of their needs.

In this chapter, we'll discuss the stages of safety marketing, which include research and analysis, objectives, and positioning. By following these stages, you'll learn to assemble and deliver a comprehensive plan for marketing safety. You'll discover new ways to motivate your people much more effectively. You'll spend less time enforcing and more time engaging. Less cop, more coach.

A word about what safety marketing is not. It isn't baseball caps, coffee mugs, or key chains. It isn't a truckload of incentives and rewards. Safety marketing isn't the hard-sell sales pitch. It's not information dumps at safety meetings. It isn't signs and banners and generic safety slogans posted around the work site. It's not about communicating rules, processes, and procedures.

WHAT SAFETY MARKETING IS

Companies spend millions on outside marketing strategies to convert readers, viewers, and visitors into buyers.

They build interactive websites and develop and hire social media staff to answer queries. They staff call centers, engage traditional media advertising, become involved in charities and sponsorships, and become more visible as part of their marketing strategy.

Marketing to inside employees involves the same functions as marketing to outside customers. It has to achieve a desired outcome. An in-house marketing campaign targets the needs and aspirations of company staff. The message has to resonate with employees, align with their personal values, and motivate them to action. What you're trying to do is convince your people to buy into your safety program and safety culture. That takes a planned professional marketing strategy.

Safety managers think they're building marketing campaigns when they download generic "Be Safe" posters from a Google search. They spend a great deal of money and effort producing posters, banners, and stickers, but abandon the effort a few months later with no discernible results. That's because there was no plan or strategy for marketing and follow-through. There was no analysis of needs.

Safety marketing is more than pulling a slogan off the Internet and expecting employees to embrace it. Catch phrases such as "Safety First" or "Think Safety" offer too little. They're too simplistic. They don't adequately engage employee values.

One of the quickest ways to fail at marketing is to market an idea that hardly anyone wants, needs, or understands. Unfortunately, safety marketing has traditionally been executed exactly this way, using cutesy Dr. Seuss-like slogans that do not connect to personal values. Here are just a few:

- Ten fingers, ten toes, if you are not careful then, who knows?
- Your head will go splat, without your hard hat.
- A spill, a slip, a hospital trip.

We need to do better. We can start by following the lead of professionals in the marketing field. In order to connect with employees through marketing, first we need to understand where our employees are coming from. Safety branding must be built around existing employee perceptions and motivations. Basic research is critical to gaining employee buy-in. It helps determine which messages will resonate and which will flop. We can never assume that we know what employees and stakeholders are thinking without bothering to find out.

DeSantis Breindel is a branding and marketing firm in

New York City with a specialty in providing marketing solutions that rally employees. They believe strongly in doing proper research before launching any campaign. The agency reminds us that by giving employees a role early on, you can build a marketing campaign they will get behind. In this way, employees assume ownership of the campaign for safety rather than seeing it as something force-fed by management. It's extremely important to get participation from as wide a cross-section of departments and job functions as possible.

Good marketing is based on research and analysis. You need to understand the mindset of your target market (employees) before you can develop a safety marketing strategy. What motivates your target audience? What mental barriers are in the way? What benefits are they looking for in safety?

Safety marketing is the link between safety and employees. Marketing captures insights about employees, connects with them, builds a strong brand for them to align with, communicates and delivers value, and creates long-term loyalty. Creating loyalty to safety is the dream of every safety manager. However, it will not happen without the support of an effective marketing strategy.

In order to build a safety marketing strategy, you'll need

to gauge employee answers to two basic questions:

- What can help me be better at what I do?
- Can I trust it?

The goal of your safety marketing strategy is to make *safety* the unequivocal answer to the first question, and *yes* the answer to the second. In marketing, it's called owning mindshare. Mindshare is a consumer's level of awareness of a product or service. If you have mindshare, you can build loyalty. Loyalty to safety reduces opposition to enforced rules and procedures.

The more you know about your target market—employees and contractors—the more successful your marketing efforts will be. You will have to make decisions about who and what benefits you must promise and deliver; these decisions are easier to make if you have done proper research. The objective is to align safety with what employees and contractors want and need.

STRATEGY: THREE STAGES OF SAFETY MARKETING

Crossrail is a 118-kilometre railway line, one of Europe's largest railway and infrastructure construction projects that runs right through the city of London. The Crossrail project has insisted that safety marketing be a central tenet of how they engage employees. Their "Have Your Say" health and safety climate survey gives employees and contractors an opportunity to make suggestions as to where the safety program needs to improve. This is called Research and Analysis, the first stage of safety marketing.

From the survey analysis, Crossrail developed an objective to engage employees and contractors to keep safety at the top of mind. This is the second stage of safety marketing: the Objective. Then they positioned a campaign that included toolbox talks and branding. This is the third stage of safety marketing: Positioning.

It is no secret that people buy into what they help create. Within your company, the most effective way to build a strong safety marketing strategy and campaign is by working with your organization's Joint Health and Safety Committee (JHSC).

Like Crossrail, your JHSC will likely already have much of the infrastructure or tools at their disposal to aid in developing a marketing strategy and campaign. This will include email addresses and survey software. They will also have people to share the responsibilities of getting it all done.

If you work for a small company without a JHSC, you

can still undertake the Three Stages of Safety Marketing. However, it will be detailed work, and you should give yourself a good amount of time to accomplish each step in the process.

In each section that follows, you will find a series of questions. Some are meant to be discussed at the committee level. Others are to be asked of employees, contractors, and those who fit the target marketing profile.

STAGE 1: RESEARCH AND ANALYSIS

Committee question: *What are we selling?* This is the starting point. You're gathering information (research) and analyzing it. You have to understand and determine the ultimate product/idea/point of view that you want your people to own for themselves. You don't yet know what your message will be. You're in the finding-out stage. What part of safety are you selling? It's not the rules or procedures; that's what training is for. Is it compliance? Is it freedom of choice? Is it teamwork?

What is the committee's point of view on safety? Work together to narrow down the committee's point of view to a single sentence. See if it captures the essence of what you believe your safety program represents. Then ask what point of view on safety you want your employees to adopt.

A word of caution here. The viewpoint that you want employees to adopt may be different from the JHSC point of view. The JHSC is likely made up of people who are already dedicated to safety. They will have a very different viewpoint than your target audience. So figure out how much of a leap you would expect your employees to take in safety when you determine what it is that you're selling.

Examples of safety committee answers to "*What are we selling?*"

- Teamwork is how we believe safety works best.
- Every person we employ is valuable, and they all need protecting.
- Safety is how we demonstrate how much we value our people.

Committee question: *What campaigns have been used in the past?* This is where you assess whether anything you have done in the past has had any success. Think of all of the posters, banners, slogans, and communications that have been used to communicate safety. Create a team effort to search out every banner, poster, and PowerPoint slide in the workplace. Take photos of them. Discuss them at your meeting. Assess their value and success.

Also, discuss what employees and contractors have communicated to you about safety campaigns. Was there improvement or decrease in safety performance with past campaigns? Why did you stop using them? What has worked, not worked, and why?

Admitting your faults in past safety marketing efforts helps clear the air. If you lost interest in a past campaign, admit it. If you were disappointed by a lack of buy-in, admit it. This will help you to create realistic expectations for a new campaign. It will also communicate that safety marketing needs to be a team effort.

Employee question: *What has the company done well in safety, and what have we not done well?* This is where you begin to involve employees. It is imperative to understand how employees feel about past safety campaigns. Will those experiences inhibit the success of a new campaign? You must have a good understanding of existing opinions and attitudes toward safety. Safety marketing needs to take into account existing attitudes and misperceptions. Your research and analysis needs to assess the collective mindset of your employees.

Assess anything communicated or done by management in the past that might undermine a new campaign initiative. Given the opportunity, employees will almost always blame management. They will point out inconsistencies from past management behavior. Be prepared for it. The existing attitudes and opinions of employees may get personal, but it's important to hear what employees feel.

The key is to avoid becoming argumentative. You are conducting a survey of opinions. You are information gathering, so your position should be neutral. Arguing perceptions only hurts your credibility. Listen and don't judge. Acknowledge all employee concerns. Take notes. Pay attention. Employees are giving you the information you need to create an effective marketing strategy that will benefit your workplace and safety program for years to come.

Employee question: *What is your current belief about safety?* You want to gauge employee thinking and feeling about safety. Which aspects of the existing safety program do they like and agree with? Which ones don't they like? Be careful not to make assumptions, but rather to gather the objective information you need.

If they don't like paperwork, or meetings, or job hazard assessments, find out why. It's the *why* questions that give you the best chance of developing a safety marketing strategy that will reap rewards.

Compile the information (research) you've gathered. Your

next step is to analyze the information. Pore over it in your JHSC meetings. If there are still gaps in the analysis, ask more questions of employees. Don't hesitate to involve even more employees. The more who participate in your informational surveys, the more who will be inclined to take ownership of the safety program they help create.

STAGE 2: OBJECTIVES

The second stage of safety marketing involves setting objectives. It requires answering the question: *What do you want the campaign to do?*

The answer is not simply to "make them be safe." A campaign cannot make people do anything. Marketing *motivates*; it *moves* people. It inspires people to voluntarily want to do something differently and to take ownership of their choices. What choices would you like employees to make? Remember, the easier the choice, the easier the decision. And also remember, your marketing campaign should reflect the analysis of your employee surveys.

What result do you want from your safety marketing? Once your people start doing something differently, what outcome or result are you hoping to achieve? For every action there is a result. Asking your people to use spotters when moving equipment in order to reduce property and vehicle damage should have a result. What result would you like to see? Is there a number, an amount, or a percentage? Be realistic. Don't expect a 100 percent drop in property damage in the first week of launching a new marketing strategy. A 30 percent decrease in property damage in the first three months of the campaign would be a significant improvement.

What philosophical and behavioral shifts do you want to motivate? Perhaps you're hoping to affect change in housekeeping measures, to institute twice-a-day tailgate meetings, or to take twenty-second focus moments. Perhaps you want to affect change in perceptions such as valuing co-workers, recognizing each other's strengths, or appreciating safety as worthwhile. Your expectations have to be specific.

Who do you want to reach with this campaign? Get specific about who you're targeting in the campaign. What job functions are you targeting? Is the campaign specifically for your truck drivers, or is it for your contractors? Is it for the whole company from the CEO to executive assistants to frontline janitorial? If it's for one or two departments, or particular crews, be specific and communicate in a way that resonates with those employees.

Developing an objective is about becoming very clear

about who you want to motivate, what you want them to do differently, and what result you want to realize.

STAGE 3: POSITIONING

The third stage of safety marketing involves positioning. This is how you communicate your objectives to your target audience. *What should be the overall tone and approach?* Will you take a soft approach or one that's more hard-hitting and sobering? How do you frame your tone? Will you encourage collaborative discussions and participation in meetings? An upbeat marketing strategy doesn't use gruesome, negative photos and stories in safety meetings. That would undermine the marketing strategy. Your tone must communicate and support your message.

Which media are you using? The traditional media, such as safety meetings, newsletters, and signage, should be expanded to include social media, web pages, awards, text messages, events and celebrations, intranet and email. An innovative marketing slogan, such as "People of Value," would be assigned a Twitter hashtag #peopleofvalue and a Twitter feed to support it.

What are your deadlines? Marketing strategies work best when they are rolled out a little at a time. A smart safety marketing strategy is time released. If you roll out everything at once, you won't leave room to build your safety brand and message over time. You don't want your campaign to lose momentum and peter out in ninety days. You want to sustain it. Assign elements to be released each quarter. Use a Gantt chart to determine which element is released at what time. Assign people to be responsible for each element. Go outside for expertise you may not have on your committee. Bring in other staffers who can help.

Can you say it in under seven words? The best marketing campaign slogans hit the mark in just a few words. This will be the toughest part for a committee. Remind your committee members to focus all messaging on the front-line employee. Focus on the major takeaways from your employee surveys. Remember: marketing motivates; it moves people. Keep your messages short. Your campaign slogan should be punchy, but most of all, make it memorable. It has to resonate and hit home.

In positioning your safety campaign, it is important to not muddy the waters. Watch out for competing or mixed messages, especially anything that is not in adherence with the safety program, such as the following:

- Long-winded safety meetings that cover too many topics at once can compete with the core safety message.
- A heavy push for increased production and speed can com-

pete with safety.

- A supervisor complaining about senior management's lack of commitment to safety can compete with the safety message.
- Personality conflicts, overbearing bosses, and fear of job loss compete with safety.
- Worst of all are the safety shortcutters who encourage others to follow their lead.

Safety must be positioned as the first choice in the employee's mind. Everything else is secondary.

Through strategic positioning, safety has the broadest reach into your target market. Positioning is about top-of-mind awareness and, ultimately, buy-in. Strong brands have top-of-mind awareness. When we think of ketchup, we think Heinz. When we think of cola, it's Coke or Pepsi. Top of mind is where safety needs to be. It's where you need to position your marketing campaign. Safety must be the first choice, not one of many competing choices.

By clear positioning, safety's presence will be everywhere your target market is. Positioning ensures that the message you put in front of employees resonates enough that they choose safety as a reflex. Safety becomes their first choice every time.

PUTTING SAFETY MARKETING TO WORK

Louisiana-Pacific Corporation has received more than sixty safety, environmental, and industry awards. They've experienced hundreds of thousands, if not millions, of hours worked without a recordable incident. Not only is safety a priority at Louisiana-Pacific, but it's also a corporate value. As a value, it doesn't have to compete for attention with other priorities.

At LP's veneer plant in Golden, British Columbia, the entire plant was shut down over two shifts for a Safety and Quality Education Day. I was invited to speak on safety leadership, and in my preparations, I discovered that LP uses all three stages of safety marketing. They connect quality of product to safety and competitiveness. LP believes that they cannot build a quality product without employing quality safety. That helps the plant to remain competitive. By being focused on quality, LP improves safety to reduce work stoppages and keeps production moving. Safety keeps costs down and the plant competitive.

LP is a safe place to work, and, as a result, LP attracts a better class of loyal, long-term, family-oriented employees. Through safety marketing, they've also reduced turnover and attrition.

In another example of safety marketing, 350 senior executives of Imperial Holdings in South Africa assembled in a conference center ballroom and individually pledged to be safe and courteous drivers. They called it the I-Pledge campaign. Imperial Holdings employs thirty-five thousand people in trucking, car rentals, tourism, and other businesses, including auto dealerships. They are one of South Africa's biggest road users.

Imperial launched a public safety marketing campaign and encouraged the general public to join with them to apply safety principles on the road. The objective of the campaign was to make driving, which is normally done almost unconsciously, conscious in the minds of employees and the general public. They wanted to engage the public in helping to make the roads they share safer for everyone. Safer roads for employees and customers means fewer accidents and less cost. Their successful marketing campaign was a win-win for the company and the public at large.

Real world examples such as these make the case for clear, strategic thinking and planning when it comes to marketing safety. By following the three-stage strategy of research and analysis, objectives, and positioning, you will join other world-class companies in promoting workplace safety effectively through innovative and creative safety marketing.

CHAPTER

9

MOTIVATION: SUPERVISORS AND SAFETY PEOPLE

Motivation is the fourth and most essential component of the M4 Method. Like a good pair of lungs, motivation breathes life into all the parts. It provides the oxygen for management, meetings, and marketing. Without motivation, management becomes an exercise in rules enforcement. Meetings go back to being dull and boring. Marketing efforts are quickly forgotten. Without motivation, all of your efforts for safety will gasp for life.

As discussed in Chapter 1, safety people generally don't know much about what motivates people. You won't find any mention of motivation in safety certification courses.

It isn't covered in safety training on the job. Rarely is it even covered in management skills training. Supervisors and safety personnel basically use rules compliance and fear of harm as motivators. We've seen how safety enforcers are more like cops than coaches. They become cops by default when they fall short of effectively motivating employees. It isn't only that they fail to motivate crews. The truth is that they fail to motivate themselves.

Yet motivation is the single most important factor in job performance. If you're not motivated, you don't get results. Motivation is fuelled by purpose, desire, and inspiration. It's what transforms safety from "have to" to "want to."

The fact is that motivation starts with the supervisor or safety person. In the same way that safety starts at home, so does motivation. The supervisor who is highly motivated to make safety *the* priority on the job will win crews over to safety. The supervisor who isn't motivated doesn't stand a chance of motivating others.

So, first things first. We begin our discussion of motivation with how you can motivate yourself. In the next chapter, we'll look at how to motivate others.

INSIDE VERSUS OUTSIDE

In order to understand how to motivate and influence others, you need to understand yourself. Is safety on the job something you feel deeply? Is it a value you hold dearly? Do you take pride in having an important role in keeping your crews safe? Do you want to make sure they can continue to feed and clothe their families and make it to a healthy retirement? If you can answer yes to these questions, then you know safety from the inside. Workplace safety is *intrinsic* to you. You know it in your heart and in your bones.

On the other hand, if what drives you about safety is pay or rewards, recognition or praise, or avoiding a penalty or firing, then safety is something external to you. It comes from the outside. It's *extrinsic*. You do it in order to receive an external reward or to avoid a negative outcome.

Intrinsically motivated people engage in safety work because it aligns with their personal values. They want to get it right not just for their own sake but also for the sakes of their co-workers and families. Maybe they've witnessed incidents on the job, or experienced them. Maybe they've seen people lose their livelihoods due to injury, and they don't want to see it happen again.

Extrinsically motivated people may be looking for recognition, a better paycheck, power and authority, a bonus, or a promotion. There's nothing terribly wrong with these things, but it's hard to translate *extrinsic* motivation to motivating others. Yet the truth is that motivating others *is* the safety person's job. That's what they're paid to do. If they want to succeed, it's time they look inside and find their own *intrinsic* motivation.

SELF-MOTIVATION

The self-motivated safety person is the one who gets results. It's just not possible to motivate employees when you're not motivated yourself. What you're selling is safety, and you've got to believe in your product; otherwise—let's face it—you're a phony, and crews will see right through you. You might win fear, but never respect. As we've seen throughout this book, fear isn't an effective motivator. People will speed down the highway when the cop isn't there.

The self-motivated supervisor has a leg up on safety. He's the one who gets buy-in from his crews. It's a whole lot easier to manage people when they know your heart is in it. People need to know you're selling the real thing, not a bill of goods. They need to know you believe it, or they won't follow through when your back is turned.

So let's say you do believe in the value of safety, or partially believe in it. Perhaps part of your motivation comes from within and part comes from the outside. Inside is heart and conviction; outside is authority and recognition. So how do you stay motivated and increase your self-motivation?

Let's compare motivation to that mobile device you use to keep track of things on the job. As a supervisor or safety person, you update the apps and the operating system on your mobile device regularly. Updates keep your device running at peak efficiency. You depend on it to help get your work done. But what about you? Are you running your own internal updates? Are you updating your mental software? Self-motivation is like an app update. It helps to remove the bugs that slow down performance. It takes some tweaking and adjusting to run at optimum efficiency.

Each day presents opportunities to recharge your motivation. Every interaction you have with others in the workplace offers a chance to make improvements. Personal interactions with co-workers, experiences on the job, and toolbox and tailgate meetings are all opportunities to connect. Every connection, if used correctly, updates your motivation. Each interaction can clarify and humanize your reasons for getting into the safety business in the first place. Even if your motivation is from the outside, the more you engage in the very human side of safety, the more motivated you become on the inside. The more you can see what safety means for crew members' lives both inside and outside of work, the more you'll embrace safety's value.

You'll find that every small improvement you make in workplace safety adds to your personal store of motivation. Safety is constantly changing and moving forward. It's your job as a supervisor or safety person to not only move forward with it, but to push it forward. If you're not doing this, then all you'll be doing is playing catch-up or falling behind. As a frontline influencer, the ball is in your court. It's up to you to bring it up court through continuous improvement. Get your people involved. Help them see the small changes that will advance the crew in their performance of safety. In that process, your motivation will grow.

PRIDE AND MOTIVATION

Every good employee wants to feel that the work they do matters. So do you. You want your work to mean something. As a safety person or supervisor, the more meaningful moments you create, the more you'll notice how you positively influence others. It's a source of pride, and when you take pride in the impact of your work, motivation naturally follows.

Pride is too often confused with arrogance and conceit. However, genuine pride in one's work is the satisfaction of a job well done. It's the pleasure of accomplishment. Pride can be shared, especially by a work crew. A crew's work is a team effort, with each one contributing their individual part. On job sites where skills and functions are interconnected, crew members depend on each other to get the job done, and to get it done safely.

A crew leader or supervisor who takes pride in his role of bringing all the parts together earns that pride. With safety as a factor, a successful day's work is also about everyone going home hale and hearty to do it all again the next day.

According to an article by Fredrickson and Branigan in *Psychology Today* (2001), "Pride ignites a positive appraisal of the self that can create feelings of optimism and wor-thiness. Unlike self-esteem, pride is triggered in response to a specific accomplishment, an achievement, an event, or a measure of performance."

Pride inspires motivation from within. A healthy dose of pride in one's own accomplishments serves to motivate the individual. And when the individual is a crew leader, that motivation extends to the crew. Motivation is infec-tious. Let's face it, leaders inspire. And safety people who take the step of becoming mentors and give up being safety cops are the ones who can truly earn the distinction of safety leaders.

Pride in successful collective effort is pride shared. It's a great motivating factor. The impact you have on safety on a daily basis compounds like interest in a good investment. It will motivate you that much more.

BEING THE EXAMPLE

Perhaps nothing will motivate you more than the example you set. If you're a parent, you probably already know this. You know if you're living up to your child's expectations. You know if your actions are deserving of respect. Of course, nobody's perfect, but it's never an excuse to stop striving.

The same values hold for your responsibilities as a safety person. Your job puts you under the spotlight. If all you have to offer are rules and regulations, fear and punishment, that's a harsh light to be under. People don't want to be pushed to do things. What they want are the tools and the relevant example of how to use them. Most employees look to their immediate supervisor for the example of "how we do things around here."

In Chapter 1, we discussed asking employees what their

vision of retirement looks like: a cabin by the lake, a motor home, time to travel, or simply being with family. When you engage crew members on the rewards of safety, they know you care. You earn their respect. When people hold you in esteem, it motivates you.

You're the person from whom they take their cues. If safety is nothing but compliance to you, it will be nothing more to them. However, if safety is a value that drives you, it will be their value too. When you own safety as a motivating value, it becomes a vision shared. The choices you make on the job become their choices. And nothing will inspire you more than the positive impact you make.

MOTIVATION RAISES THE BAR

Safety is all about being focused in the moment. It's about what's at the local level right in front of you. You measure yourself and your success by your own progress. Measuring your own safety efforts against the industry average doesn't cut it. The fact is that industry standards, as we've already seen, are too low. Average performance isn't all that inspiring. It's not exceptional. It's not best in class, and it's certainly not something that builds personal or internal motivation. Motivation is what raises the bar.

As a frontline supervisor or safety person, your goal is

to embody that higher standard. Safety as a focus and mission is a more powerful internal driver than mere compliance. Compliance with industry minimums breeds only mediocrity. As Andrew Carnegie once said, "People who are unable to motivate themselves must be content with mediocrity, no matter how impressive their other talents."

Safety allows us to stay on purpose despite the distractions. A motivated safety person is employee focused and crew focused the way a good coach is player focused and team focused. The coach who aspires to excellence *breeds* excellence, not mediocrity.

So don't measure your crew's safety performance against other crews or industry averages. Instead, set your own high standards and attainable goals. High standards are great motivators. The higher your reach, the farther you'll go and take your crew with you. Mediocrity and motivation don't mix. In fact, they work against each other. Either you motivate for safety, or you'll wind up chasing mediocrity.

As we've seen, it's the highly motivated supervisors and safety people who have the most potential for influence. In the next chapter, we'll take a closer look at how to translate that self-motivation into motivating others.

MOTIVATION: HOW TO INFLUENCE YOUR EMPLOYEES

This may be the scariest chapter in the book for frontline supervisors and safety people. The thought of motivating employees may conjure up images of giving inspiring speeches full of "live your dreams" and "you can do it" platitudes. Relax, we'll be looking at much more effective ways to motivate your crews.

Let's start by reviewing how we usually get motivation all wrong. Throughout this book, I've been calling attention to the fact that enforcing rules doesn't really motivate people. Neither does nagging and fault-finding. But we sure do a lot of both. The bottom line is that the inexperienced supervisor who doesn't know how to motivate and develop individuals on the job ultimately has a harder time getting the job done.

Gallup surveys peg the employee disengagement rate at over 70 percent. When motivation is missing, so is engagement. Enforcing rules and procedures is an exercise in policing that fails to motivate in the right way. All it does is motivate employees to avoid getting caught. Also, buying employees off with cash incentives for following safety rules doesn't work long term. Instilling fear with gruesome photos and gut-wrenching stories of "don't do what I did" also fails to motivate.

None of these tactics builds positive motivation. As discussed in the last chapter, you have to be motivated to motivate others. It's time to build on that. If you've come this far in the book, and you've taken Chapter 9 to heart, you're ready to tackle the challenge of employee motivation.

THE MOTIVATION WAVE

There's no such thing as being unmotivated. Employees who seem unmotivated are simply not motivated by the same things you are. Maybe they're motivated by nothing more than short-term plans such as going out with the boys after work. Maybe they're focused on retirement. Maybe they have dreams of starting a small business. People are motivated by different things. Whatever those things are, they may not include safety. As things stand now, safety may not be as important to them as it is to you.

BJ Fogg, professor of human performance and founder of the Persuasive Technology Lab at Stanford University, framed the concept of competing motivations. In competing motivations, our best efforts to practice safety can be thwarted because something else more important comes along. It could be a family emergency or bowling night, whatever competes for employee attention. It isn't that people aren't motivated. They just have competing motivations outside of following safety rules and procedures.

Fogg takes competing motivations to the next level when he describes what he calls a "motivation wave." As he explains it, motivation comes in high and then drops low, like a wave on the ocean. Fogg suggests that it's best to use the high point of the motivation wave to accomplish the hard things. Then, when motivation levels drop, the hard things have already been done. As he points out, "You don't need motivation to do the easy things."

Take healthy eating as an example. Having healthy snacks instead of junk food is easier when the healthy snacks are prepared beforehand and ready to eat. That's the high point of the wave. You've gone to the grocery and returned home with a bag full of fresh fruits and vegetables. At that moment, your motivation to eat well is high. While you still feel motivated, it's best to wash and prepare the healthy snacks right away and put them in containers in the fridge. Then, when your motivation drops, your healthy snacks have already been prepared and are ready to eat. This is the dynamic of using the high point of the motivation wave to set up a structure for when your motivation wave drops low.

This same dynamic can be transferred to safety. Do the hard things and create a structure when motivation is high. Then, when motivation drops, the structure will already be in place to carry through. Use the motivation wave to encourage good habits to grow naturally. However, don't expect employees to make huge leaps in changing behavior. Instead, focus on incremental changes, small things that add up.

WHEN APATHY IS IN THE WAY

The enemy of safety is apathy. To win the motivation battle, engage the enemy. Put down the rule book and have a conversation. Get employees to feel something real about safety. Equate safety with pride—pride in a job well done. Start by asking questions, drawing people out about how they view safety and what motivates or doesn't motivate them on the job.

Jim Lundrigan, operations superintendent at Sudbury Integrated Nickel Operations, knows what it's like to be motivated by a supervisor. He spent ten years as a miner before moving into a supervisory position of his own.

"Before I got into supervision," Lundrigan says, "having a manager stop to ask me what I was doing or how my family was or how my day was going was always very special to me. It would put a little spring in my step, and I would try harder to be better at my job."

Lundrigan took that example and made it his own. He knows from experience that every employee has a sense of pride. He believes that if a supervisor damages that pride, they will lose employee respect. What's left is only apathy. A supervisor or safety person who barks orders may get the job done, but an employee won't be motivated to put himself out or do anything extra for that supervisor.

Smart supervisors tap into employees' natural pride in their work. They build employees up by calling attention to the contribution they make on the job. They accentuate the positive to light a fire of inspiration and motivation to get the job done safely and efficiently. In this way, smart supervisors cut through apathy.

GOING ONE-ON-ONE

To understand what motivates employees, first, you have to understand their fears—the biggest of which is that people are afraid to look foolish or stupid. They may not ask questions for fear of being teased or singled out. This is especially true in safety meetings. That's why one-on-one time with an employee is so important.

One-on-one is where you can get past the excuses and apathy that interfere with an employee's motivation. One-on-one is where you can tell them that you care about them and their contribution. You can point out their value to the team. You can communicate that you depend on their strengths to help the team achieve. One-on-one sessions can help massage an individual's natural pride and inspire motivation.

HOW TO MOTIVATE OTHERS

There are no tricks, secrets, tactics, or manipulations for turning on an employee's motivation. It's done organically by doing the little things right. Showing people that they matter and that you respect them is the foundation of establishing employee motivation. Motivating others is fundamentally about genuine respect and caring.

There are eight things you can do to lay the foundation for influencing motivation:

1. *Be on time.* Show up for meetings and start them on time. Don't make people sit around while you fumble with audiovisuals or wait for stragglers. Honor their time. Have everything tested and ready before they arrive at the meeting. Respect inspires motivation.

2. *Be available.* Let people know that you're there as a resource. Ask for questions and comments. Listen intently and hear people out. Don't interrupt. Make eye contact. Focus your attention on them. Don't check your phone. Don't rush anyone. Make the time you spend with your crews the most important thing in your workday.

3. *Be nice.* Don't use being busy as an excuse for being rude. Saying good morning takes little effort, but counts for a lot. Calling employees by name is easy too. Always return phone calls and emails in a timely manner. Bring donuts to the lunchroom. Congratulate people on their achievements. Ask about their families. It's motivating to work in a place where people are nice to each other. You set the standard.

4. *Let them know they matter.* On nearly every single list of what motivates employees is letting them know they matter. People respond positively to praise. Show them that their

contribution makes the company better. Employees need to know that they matter to the company.

5. *Lift them up.* When employees feel that you are proud to have them on your team, they are motivated to prove you right. The best coaches have mastered this skill. Celebrate personal best performances. Encourage each crew member to achieve their best every day.

6. *Walk the talk.* Supervisors and safety personnel set the standard for others to follow. Arrive early. Keep your promises. Practice courtesy and respect. Most of all, be motivated. Positive employees follow a positive supervisor who makes safety a positive experience.

7. *Be clear in your expectations.* Don't just present information to an employee and hope that it changes a behavior. Set clear expectations for safety performance and follow up with encouragement.

8. *Be positive.* A research paper titled "The Role of Positivity and Connectivity in the Performance of Business Teams" by Emily Heaphy and Marcial Losada found that positive feedback motivates people to continue doing well and with more vigor, determination, and creativity than negative comments. The most effective teams received more than five positive comments to one negative comment. The research showed that a boss's positive comments constructively impact employee motivation.

MOTIVATING EMPLOYEES WHO ARE
IN IT FOR THE PAYCHECK

One of the questions I'm asked most often is, "How do you motivate someone who is only working for the paycheck?"

The answer is simpler than you may think. Stop forcing them to work for the paycheck.

Let me explain.

TINYpulse is a performance management tool used for ongoing surveys of employee engagement and organizational culture. In 2014, the tool was used to analyze some two hundred thousand anonymous survey responses. The findings were revealing about employee motivation and engagement.

Slightly more than half of the employees surveyed said they were satisfied with the performance of their direct supervisor. Half were not. The reason? Skill levels. In most instances, the best performer in a frontline job gets tapped for the supervisor's position. The person who studied to become a crane operator, an electrician, or a mill operator is now thrust into a supervisor's role. You may have studied to become a steamfitter, but you didn't study to become a supervisor. Your steamfitter's ticket and skills don't do much good in motivating for safety.

Supervisors without effective supervisory skills just aren't prepared, so they revert to what they know and what they've experienced: enforcing rules. They don't have the training for creative thinking about safety. When employees feel that authority is being hung over their heads, or that they'll be replaced if they don't shape up, you can kiss motivation goodbye. All that remains is alienation and fear. At that point, all an employee is working for is a paycheck, because it puts food on the table. That becomes their only motivation.

OPPORTUNITY FOR GROWTH

The survey also found that two out of three employees don't feel that their job offers any opportunity for growth or advancement. Most employees aren't empowered to take on additional responsibilities. If employees can't see how to progress in their careers, they feel stuck. So they resign themselves to coming to work only for the paycheck.

To shake them out of that thinking, supervisors and safety people have to do two things. First, they need to improve their supervisory skills. Reading this book is a good start. There are also workshops and seminars that can help to improve management skills.

The second thing to do is to engage and empower employees to play a bigger role in improving safety. Give them more ownership and accountability. Organize creative problem-solving sessions on recurring issues. Brainstorm to build better teams. It's your job to build morale. Let employees know that their input is needed and matters greatly. Give them reasons to feel that they have much to offer. Build opportunities for growth on your team.

PEER INFLUENCE

Peers are an employee's largest influence. Fellow workers drive an employee's motivation to perform better, more so than a supervisor or even money. Crew culture is first in the eyes of employees. People naturally want to fit in. If their peers are working, employees will get busy working. If their peers are focused on safety, other employees will focus on safety.

Teammates motivate other teammates to go the extra mile. If the crew is willing to look out for each other, every member of the crew will follow that culture. There's an obligation to fit in. So if you want to raise the motivation of an employee, there is no better motivator than a motivated crew of peers.

Just under half of employees (44 percent) will give a fellow worker recognition. When employees are happy and motivated, that number rises to 58 percent. Only 18 percent of the least happy and least motivated employees bother to recognize a fellow team member for work well done.

Since peer influence is the biggest driver of motivation, if a majority of crew members are motivated, others will be motivated too. It means more to an employee to be recognized for safe work by a fellow worker than by a supervisor or safety person.

Smart supervisors will work with peer influence to build a crew culture around safety. Encourage peer recognition at toolbox and tailgate meetings, formal safety meetings, on the job site, and on the shop floor.

FEELING VALUED

The biggest surprise in the TINYpulse survey was that only one in five employees felt valued at work. Four out of five did not. Seventy-nine percent feel either undervalued or marginally valued. When no one tells you that you're doing a good job, even informally, it's hard to feel any other way. Feeling undervalued directly affects motivation. When employees aren't appreciated for their work, they resign themselves to working for the paycheck and nothing else.

Some employees will even leave a good paying job for

one that pays less money at a company where they feel more valued. Simply put, money alone doesn't improve motivation. In fact, money doesn't even crack the top-five reasons people are motivated to give their best at work.

WHAT MOTIVATES EMPLOYEES TO EXCEL AND GO THE EXTRA MILE?

According to the TINYpulse survey (https://www.tiny-pulse.com), the highest-rated factors that motivate employees to excel in their work and go the extra mile for their employer include

1. Camaraderie, peer motivation
2. Intrinsic desire to do a good job
3. Feeling encouraged and recognized
4. Having a real impact
5. Growing professionally
6. Meeting client/customer needs
7. Money and benefits
8. Positive supervisor/management
9. Belief in the company
10. Other

Notice where money and benefits are in the list: seventh. When employees tell you that they're only in it for the paycheck, it's because the six highest motivators are missing from their work, along with the bottom three. No amount of money can make up for a lack of other motivating factors. As a supervisor and safety person, you have to change that.

There are five things you can do immediately to build motivation and the desire to excel, and to get your employees beyond focusing on the paycheck:

1. *Gain the skills you need.* You may already have management skills, or maybe you don't. There's always room for growth. So plug the holes in your skill set. Your success at employee motivation is directly tied to your managerial strengths. You wouldn't let an untrained worker operate a D9 Cat without the required skills. So attend management seminars and workshops in your area. See if your employer will cover the cost of professional development. Talk with other managers at your company and learn which management tools work best for them. Build your own network of advisors who have the experience you need. Look back at Chapter 9 and work to increase your own motivation. It will inspire you to learn and experiment more. That inspiration and motivation will also rub off on your crew.

2. *Mentoring and coaching.* Think about the best boss and the worst boss you've ever had. Model your supervisory style on the best example. You are the role model for safety and self-directed motivation. You set the positive tone for

safety. As we've seen throughout this book, a true safety leader is a mentor and a coach, not a cop. A safety leader doesn't manipulate. A safety leader inspires. Provide the support, enthusiasm, and concrete strategies to make your team a winning team. Coach each of your players to bring their best skills into play for safety on the job. Help each of them reach and excel at their potential. Take their input and hear them out. Engage your players as a team.

3. *Tell them how much they matter.* People are motivated to embrace safety when they know you care about them. One of the greatest sources of motivation is knowing you matter. When people feel valued, they are in it for more than the money. Let each crew member know that they matter to you. Let them know that you count on them as an integral part of the team.

4. *Encourage employees to recognize each other.* Employees are motivated by their crew. It's their community. When the crew is tight and recognizes individual effort, everyone feels more encouraged. Develop a recognition vehicle, such as safety high-fives. Encourage constructive communication at toolbox and tailgate meetings. Ask, "What are we doing right?" "Who took charge today?" "Who embraced safety as a value?"

5. *Create more leaders.* The purpose of a fruit tree isn't to grow fruit—it's to grow other trees. Grow and groom your eventual replacement. Create the potential for leaders among your crew. You'll want a smooth transition to another

capable person when you're ready to move on to another position or into retirement. Be a leader who creates leaders. It will be your legacy.

SENIOR MANAGEMENT AND SAFETY MOTIVATION

You may notice an absence of discussion about senior management in the context of motivation. That's because executives have nothing to do with frontline motivation. Senior management lacks the personal contact to reach out to frontline crews. That's where supervisors come into play. Leadership on the front lines begins with supervisors and extends to the crew.

A good supervisor can keep a team together and motivated even when senior management lacks a commitment to safety. You're the person who employees interact with each day. Senior management may be responsible for the health of the forest, but supervisors and safety people are responsible for the health of each individual tree.

MOTIVATION AS A SUPERPOWER

As a supervisor or safety person, it's your responsibility to help your people build strong roots in safety. You're the one who models motivation and gets the crew on board to do safety right. You're the person your crew depends on to set the motivational example.

If you had the power to motivate every employee you came into contact with, you'd be the most valuable person in the company. So be that person. Safety leaders aren't born; they work for it. If you do the hard lifting, your crew will be happy to lift along with you.

LAUNCH: BECOME A SAFETY LEADER

The one lesson I hope everyone learns from this book is that being a manager, a supervisor, or safety person doesn't make you a leader. Management is paid. Leadership is earned. Safety leaders inspire their teams by rolling up their sleeves and engaging one-on-one. They educate their crews by demonstrating practical applications of processes and procedures. It's an approach that builds confidence and mutual support. Leaders embrace a strategy of coaching and mentoring crews for better performance. Leadership requires an attitude that most of us have to work to develop.

The problem is that management has tried to steal the term "leadership" and make it theirs. A manager doesn't automatically become a leader by virtue of their position or title. Your crew certainly won't view you as a leader until you demonstrate by action, behavior, and temperament that you are one. Leadership is selfless. Management is merely territorial. Leaders model safety on and off the job. They're the ones others follow instinctively. They inspire their people to buy into safety as a personal value.

RESPONSIBILITIES OF A LEADER

Take a look at safety person profiles on LinkedIn, and you'll notice a worrying trend. You'll see lists of responsibilities such as the following:

- Responsible for the company's health, safety, and compliance issuances
- Deliver and maintain safety in the workplace
- Identify hazardous conditions and practices in the workplace

Sounds pretty good on the surface, but it isn't the profile of a safety leader; it's a job description. Any safety person would be expected to perform these exact same tasks. And if that's all a safety person can claim, then they're completely replaceable. There's nothing that differentiates them as a leader.

Anyone can be a supervisor or a safety person, but very

few are truly effective at it. Being effective means you've embraced the skills and values necessary to positively influence your fellow workers and employees. Mediocre supervisors and safety people use authority because it's all they have. To have real influence, however, requires having the trust and acceptance of your people.

SAFETY AS A MISSION

A mission isn't a job title, a position, or a list of responsibilities. It's a calling. It's a sense of purpose. It defines who you are and how you choose to live and work. A mission is a way to focus your energy, actions, behaviors, and decisions. It gives you a reason to get out of bed in the morning beyond just doing your job.

To define your mission, try answering the following questions:

- Why do you choose to be in safety?
- Why is it important to you?
- How can you make a difference in workplace safety?

Think about each question and write down your answers. Then finish this sentence: "My mission in safety is…" Keep it short; write it down, and carry it in your wallet. Look at it every day and see if it still feels right. See if it helps you stay focused on what you want to accomplish as a safety supervisor. Does it inspire you? Does it change your orientation to your work? Does it motivate you? Does it provide focus for how you work with your crew?

QUANTIFY YOUR EFFECTIVENESS

As you make real gains in safety improvement at your company, you'll eventually need to measure your effectiveness. Safety has a direct effect on the business side of an organization. Strong and effective safety leaders can positively affect profit and loss, revenue, turnover, and attrition.

In order to quantitatively assess both your organizational and personal effectiveness as a safety person, ask yourself the following four questions:

1. *Am I helping to drive revenue to my employer?* In other words, is the safety program performing well enough that it is helping the company to secure more business and new clients? Become aware of and quantify how your company makes more money because of the performance of the safety program. Safety leaders comprehend the value of safety and can argue it with proof.

2. *Am I helping the company to reduce costs and maximize profit?* A successful safety program will do both of these. Staff

turnover and attrition incur hard costs for a company. Insurance premiums are reduced for good safety performance. These are metrics that can be measured. If you are going to argue that safety needs to be integrated into organizational departments, these arguments are best made at the level of profit and loss. This will be beneficial when you approach senior management to boost investment in safety.

3. *Are my co-workers improving both personally and professionally as a result of working with me?* Do you help your co-workers to be more effective? Loyal employees who commit to continuous self-improvement also help a company reduce costs. Safety is a driver that contributes to more satisfied employees and a positive work environment and corporate culture.

4. *Will I be missed when I am gone?* Through either retirement or moving on to a new job, at some point, you will split company with your employer. Will you be missed when you leave? Or will it be easy to find your replacement?

True safety leaders are not just focused on rules and compliance. They help organizations integrate safety into the everyday operations of every department. Safety has an impact on a company's bottom line. The valued, irreplaceable safety leader will always be in high demand.

PUSH YOUR BOUNDARIES

There's no such thing as the status quo. You're either moving forward or you're falling behind. Everything is in a constant state of motion. In order to remain relevant and effective in our work, we all need to improve. We do this by learning to push our own boundaries.

When you raise the bar for yourself, and reach it, then and only then can you raise the bar for your team. If you only measure yourself against the industry average, you'll barely manage to keep pace. Your team will simply not be responsive. Teams are only motivated by highly motivated leaders. It's as simple as that. Minimum standards for safety are a trap of mediocrity. It only dumbs down the team. Instead, push to surpass the average in safety performance.

Aim for the gold. Championships aren't won by mediocre performers. To become a leader, you're going to have to stretch yourself. To motivate as a mentor and coach, aim high and your crew will support you.

COMMUNICATE WITH PURPOSE

Information in the digital age is coming at us faster, in larger volume, and from more sources all the time. In fact, it's a constant interruption. Our brains are working harder than ever to process it all, to separate out what's relevant, and to ignore what isn't. Crew members and employees have a lot of information coming at them too. What seems like a straightforward communication to you can be perceived as an interruption by your crews. They're already busy with their work. Their hands and minds are already engaged. They have to sift through the information to see how it applies to the task at hand, or another task.

So when we communicate with our crews about safety, we can't expect things to stick the first time. We can't expect to just say something once and know it's been completely understood. We can't expect results without patiently following up. This is why I prefer face-to-face communication. It's personal and more immediate. Respectful eye-to-eye conversation shows you care. It provides greater assurance that a crew member is mentally present and listening. It's a chance for real interaction, response, and dialogue. It's the kind of dynamic that helps new information stick.

Communicate clearly and on point. What is it that your people need to know to do their jobs better? Not everyone understands the same message the same way. For example, a simple word such as "dog" conjures up different images: different breeds, colors, and sizes of dogs.

When you say "safety," everyone sees something different as well.

So you've got to be purposeful in your communication. Be specific. Get to know your people and their lives. Find out what's important to them. One person is a bodybuilder, another is an amateur chef, another is adding a deck on his house. When you talk to them, make them feel that they're the most important person in the world. When they feel that kind of attention, they'll offer the same attention in return.

FOCUS ON INDIVIDUAL BUY-INS

As appealing as it seems to get all of your employees to buy in at the same time, it won't happen. Each employee needs to find their own win-win before, that means you'll have to fight for their hearts and minds one at a time.

The best strategy is to start with the low-hanging fruit. Find people who have already shown the ability to make good decisions. Ask for their commitment to help build a better crew for everyone. Then once you have a few of these people in your corner, the rest of the crew will take notice, and the momentum will build.

Express things in positive ways. Not getting injured isn't

enough motivation. Working effectively and safely to build a future for oneself and one's family is. People only buy what they see and like. Doing one's part and working efficiently as a team to get a job done is something individuals can buy into.

Speak regularly with your people and keep the conversation going over time. Always ask for input and always follow up. People want to know they are being heard and listened to. Buy into them and they'll buy into you and your safety message.

COURTESY IS RESPECT

If you can hold a door open for a senior citizen, then you can extend courtesy. Courteous people are impossible to ignore. Courtesy makes people feel comfortable. It makes them want to do business with you.

Courtesy is inherently unselfish. It fosters respect. When you give it, you receive it, because you've earned it. Basic courtesy can change workplace dynamics for the better. Try it and you'll see. Courtesy in words and deeds builds upon itself and spreads.

ADVICE FROM JIM LUNDRIGAN, OPERATIONS SUPERINTENDENT

. . . . . .

"Art Gardiner was one of the nicest supervisors I ever had. He could come in and give you shit in such a nice way that after he left, you felt bad for disappointing him. It was like having your grandpa look at you and roll his eyes and tell you he was disappointed. That's the stuff that rips your heart out. Art was the kind of guy you just wanted to make happy and proud. He never yelled. Never raised his voice. He would just look you right in the eye and express his quiet disappointment. That would cut a guy more than getting yelled at or getting a written warning. Those are the guys that you respect. So I've tried to fashion my own management style after Art's.

"It's a challenge for a new supervisor. You've got to find your voice. And it takes at least a couple of years for you to figure out who you want to be as a supervisor. You're going to make mistakes. First, you're going to be the hard-nosed, zero-tolerance, no-bending-the-rules kind of guy. Then you may flip over to buddy-buddy with the crew just so you can get the work done. Maybe you overlook things. Maybe you allow too many guys off at any one time. Until you find the path where you're comfortable, and you can still enforce the rules, still get respect from the guys, and still manage to keep your sanity, you'll need to find your own voice. It takes a good couple of years for a good supervisor to figure that out."

GIVE EMPLOYEES SKILLS TO MAKE BETTER DECISIONS

The success of a safety program isn't determined by TRIF rates or LTI scores. It also isn't determined by rules enforcement. The success of a safety program is determined by what happens when employees return home each day. It's about whether or not they embrace safety in their personal lives.

This is why the real buy-in to safety happens when safety is embraced as a personal value. Until employees embrace the safety program, they're only going to tolerate safety rules and will need to be policed into compliance. The only way this can change is with mentoring, coaching, and building an ethic of teamwork focused on safety culture.

What we want to do as supervisors and safety people is encourage personal action and decision-making ability. You want to empower your crew members to make good decisions. What you need on your team are influencers: highly motivated people who motivate others. You find those people and influence those people by being one yourself.

CARE FOR YOUR PEOPLE'S SAFETY

Are you really concerned with your people's safety or are you more focused on just getting them to follow the rules?

If you're concerned with keeping your people safe, then you genuinely care that no one on your crew gets hurt. The way to communicate this is to change the conversation from admonishment to caring.

The message you want to convey is, "You're valuable to me and the rest of the crew. I don't want you to get hurt or miss work because, quite frankly, we need you here."

Once they know you care about them as individuals and as important contributors to the team, they'll start to care more about safety. It also shifts the way they see you. You go from being a safety cop to being a coach your people hold in esteem.

VIEW YOUR EMPLOYEES AS GOOD PEOPLE

If you can't trust your people to do the right thing when your back is turned, then you don't really have their respect. You should take it personally. The change you need to make is in yourself. Because before you can assert influence, you have to earn respect. It's time to reflect on the quality of your leadership.

A supervisor who's enforcing the rules, punishing, admonishing, or mocking behaviors only alienates their crew. It sucks to work for a bad boss.

Remember Chapter 4 and "People View?" As a supervisor or safety person, you can either elevate or drag your people down. You can use your influence to improve performance, or you can follow your people around waiting for an opportunity to scold them. How you conduct yourself on the job site will vary greatly based on how you view your position and how you view the people you work with. You need to start seeing your crew members as the good people they are. They have families, children, and close friends. View them this way, and the way you supervise them will naturally change.

Understand and get to know your crews as the kind of people who'll help a neighbor out. See them as the kind of people with the skills and capacity to respond to a natural disaster, because that's the kind of people they are. These are the people who can be better motivated by positive goals than by threats. They're the kind of people who know how to pull together to get a job done. Your job is to help them see how to do it safely for everyone's sake. It's your job to create an atmosphere in which people rally together to get the job done efficiently, then go home safe.

Your people want to feel proud of the work they do. They want to be admired and appreciated for their specialized skills. Your job is to provide them with the tools, the knowledge, and the motivation to make good decisions every day.

WHAT IS A SAFETY-MINDED ORGANIZATION?

In most cases, safety has been viewed as an add-on to existing departments. Organizations create all of their processes and then tack on safety rules at the end. This creates an internal conflict between production and safety. However, the truth is that the best production is safe production.

The fastest way for a company to waste money and resources is to have a serious safety incident. When this happens, people just stand around burning time on the clock to determine the cause. Materials are en route and on schedule, money is being spent, but production is halted.

Smart companies know that the way to make money and maintain high production is to have employees highly motivated around safety. This turns safety from an add-on to an organizational principle.

In safety-minded organizations, everything is planned, designed, and accomplished with safety as an underlying value. This involves a people-oriented viewpoint that sees productivity and safety as not only compatible but also necessary for success.

As a supervisor, you don't have control over corporate

identity, but you can control your crew culture. You can lead an exceptional crew and set a company standard by example. You can motivate exceptional people who do exceptional work. Nobody can take that away from you. Instead, they'll want to know how you do it. It's how you can have your greatest influence at work and extend your legacy.

MAKING A DIFFERENCE AT SAFETY MEETINGS

Supervisors often feel the pressure to rely on authority to get people to do things. However, as we've seen, authority isn't a motivator. Motivation is tied to your influence as a leader. Showing people that they matter and that you respect them is where motivation starts.

When you truly believe that your people are important, you'll work hard to involve them in your safety meetings. Supervisors and safety people who rely on PowerPoint presentations and numbers are just being lazy, and all they get is lazy attention. Uninspiring meetings yield uninspired results.

Dare to be different. Put away the bullet points and ask questions. What does accountability mean to you? Have a group discussion about what it means to your crew members. Try other questions to open discussion:

- What's the best thing we do in safety at this company?
- What's preventing us from working safely today?

Questions like these engage thought and participation. That's what you want. You need your people to be actively thinking about safety. Talk about safety in the same way a sports team talks about covering the field in a game. When you get your crew to concentrate on covering each other's backs on the job, you're building a solid safety culture. For a meeting to be engaging, everyone needs to feel they're important. Everyone needs to know they have a voice and will be heard.

IGNORE TROLLS AND CRITICS

There are a lot of armchair critics who are more than willing to tell you what you're doing wrong. That's because criticizing is easier than doing things. Some will tell you that you shouldn't get close to your employees, in case you need to report them.

The best way to deal with critics is to ignore them. Otherwise, they'll suck up your oxygen. They'll tear you down to elevate themselves. Don't give them the opening or opportunity. Stick to your plan.

Maintain your people-centered focus on safety. Build your

own community of like-minded co-workers and managers. Offer support to others who want to improve. By keeping your attention where it should be, and building a culture of safety among your crew and employees, you'll silence the critics.

YOU CAN'T BUILD QUALITY WITHOUT SAFETY

PENTA Building Group prides itself on high-end finishes for projects that include hotels and casinos on the Las Vegas strip, Zappos world headquarters, and the new T-Mobile Arena, home of the new National Hockey League expansion franchise. When safety manager Rodd Weber took me on a tour of their corporate offices, I was struck by all the framed photos of PENTA projects. In my conversation with Rodd and operations manager Steve Jones, it became apparent that the company's high standard of work relates directly to their high standard for safety.

As I noted in Chapter 1, safety is a foundational value at PENTA, and the difference this has made in their safety record is remarkable. In 2015, among their six-hundred-plus employees in Las Vegas, there were only four recorded incidents and all were minor.

The fact is that there's no way to pride yourself on the quality of your work without also having a high-quality safety program. You can't view yourself as a quality leader while employing low-quality strategies for safety. It just doesn't work that way.

Building quality on the job starts at home, with you personally. Invest time and energy in yourself, including your own learning and health. If you've read this far, you're already partway there. As you take the lessons of this book to heart and work the ropes of safety as an engaged coach, you'll gain influence among your people for improving quality together.

ARE YOU READY?

We've all had someone we've wanted to emulate. These are the people who've earned our trust, admiration, and respect. They've influenced our career paths and the life decisions we've made.

This is the person you should be for your crew—the person your co-workers can look up to, the person your kids and spouse depend on, and the person your parents are proud to have raised. This type of person leaves the strongest, longest-lasting legacy and raises the bar of expectations for what's possible. That person is you, or could be you. It's within your power and reach.

Throughout this book, I've presented a method to help you get there. It's the foundation of your game plan. I hope you use it.

In the introduction, we talked about the powerful effect that a simple phrase can have on someone: "I really want you to be around for a long, long time." That's how we feel about the most important people in our lives: our families and our friends. It's also the basis for how we need to feel about our crews and co-workers on the job, especially on work sites where a single misstep or lapse of attention can affect the safety of not only ourselves but also the people around us.

To get there you'll need to gain the respect of your people and understand your role in motivating and influencing them. You'll need to change your focus from rules enforcement to coaching, mentoring, and building mutual respect.

You'll also need to embrace the new safety model that I introduced in Chapter 2. Remember, it's a people-centered method that picks up where ordinary safety programs leave off. It goes beyond procedures, processes, programming, and production to embrace the most valuable component of safety: people.

It's a method of integrating people into the strategy of

building solid safety programs. It directly helps supervisors and safety people build relationships that support safety from the ground up. As you'll recall, this M4 Method combines four critical components to achieve the next level of safety in your workplace:

- Management
- Meetings
- Marketing
- Motivation

It's not designed to replace your existing safety program. The M4 Method is designed to augment your program by focusing specifically on people. Your people are your greatest asset. They are the critical component of building a culture of safety.

In my own work, I'm a management consultant with a focus on safety in the workplace. My company, ZeroSpeak Corporation, is a communications firm in the area of safety management. We help our clients connect with the hearts and minds of their employees to improve safety and wellness. I'm an avid blogger and keynote speaker dedicated to helping frontline supervisors and employees, as well as managers and executives, retool their company's safety culture. Feel free to contact me at www.kevburns.com to continue the conversation.

You now have the tools to motivate your people and build a safer workplace, if you choose to. But will you? Can you take the steps to become a safety leader?

I think you can. It's the same way I improved my personal and professional life. It's how I changed my whole outlook on safety. All because someone I cared deeply about told me, "I really want you to be around for a long, long time."

Traits of Safety Leadership

ABOUT THE AUTHOR

KEVIN BURNS is a management consultant and an international thought leader and speaker based in Calgary, Alberta, Canada. He is the CEO of ZeroSpeak Corporation and principal consultant at M4 Management Consultants. Kevin originated the M4 Method of interpersonal management skills development for supervisory and safety personnel, and he has authored ten books on human performance and safety. He speaks at local, national, and international safety conferences, as well as corporate safety events and meetings.

As a workplace safety management consultant, Kevin helps companies and organizations integrate people-centered principles into their safety programs. He works with senior executives, middle managers, supervisors, safety personnel, frontline employees, and contractors to establish safety as a personal and company-wide value. His innovative approach, called PeopleWork, guides organizations and individuals through practical steps that promote productive, safe, and healthy lives.